RANDALL K. HOLMES, M.D., Ph.D.

Trafficking
of
Bacterial Toxins

Editor

Catharine B. Saelinger, Ph.D.

Professor of
Molecular Genetics, Biochemistry, and Microbiology
University of Cincinnati
Cincinnati, Ohio

CRC Press, Inc.
Boca Raton, Florida

Library of Congress

Trafficking of bacterial toxins / editor, Catharine B. Saelinger.
    p. cm.
  Bibliography: p.
  Includes index.
  ISBN 0-8493-4787-4
  1. Bacterial toxins. 2. Ricin. I. Saelinger, Catharine B.,
1943- .
  QP632.B3T73 1990
615.9′5299—dc19
                                         89-645
                                           CIP

Direct all inquiries to CRC Press, Inc., 2000 Corporate Blvd., N.W., Boca Raton, Florida 33431.

© 1990 by CRC Press, Inc.

International Standard Book Number 0-8493-4787-4

Library of Congress Card Number 89-645
Printed in the United States

PREFACE

The pathway followed by a ligand within a mammalian cell determines the expression of its biological activity. A growing number of ligands have been shown to enter cells by receptor-mediated endocytosis via coated pits and to be delivered to acidic prelysosomal compartments from which they are directed to the next station on their journey. Plant and bacterial toxins appear to have usurped certain of the normal pathways of the cell for their own use. Coated pit-mediated entry has been described for some toxins, but may not represent a universally traveled path. While movement through several membrane-bound vesicles is a common phenomenon, the site at which the toxin enters the cytosol to interact with its substrate differs with the toxic protein being studied. In some cases, acidic compartments are involved, and in other cases, basic compartments participate in the trafficking process.

The articles on the following pages address many of these areas. In some instances, definitive evidence is available on many of the steps in the toxin's journey. In other instances, only minimal information is available. In these circumstances, the speculations of the authors should prove interesting to the reader.

THE EDITOR

Catharine B. Saelinger, Ph.D., is Professor of Microbiology and Molecular Genetics in the College of Medicine at the University of Cincinnati.

Dr. Saelinger received her A.B. degree from Smith College, Northampton, Massachusetts in 1965. She obtained her M.S. and Ph.D. degrees in 1967 and 1970, respectively, from the Department of Microbiology, University of Cincinnati. After doing postdoctoral work at the University of Louisville School of Medicine, she was appointed an Assistant Professor of Microbiology at the University of Cincinnati. She became an Associate Professor of Microbiology and Molecular Genetics in 1982, and Professor in 1987.

Dr. Saelinger is a member of the American Society for Microbiology, the American Society for Cell Biology, the American Association for the Advancement of Science, and the honorary society, Sigma Xi, and is a Fellow of the American Academy of Microbiology.

She has been the recipient of research grants from the National Institutes of Health. She is the author of over 50 papers. Her current research interests are in virulence factors associated with *Pseudomonas aeruginosa* and in determining the intracellular trafficking and processing of bacterial toxins by mammalian cells.

CONTRIBUTORS

Viloya S. Allured
Department of Chemistry and
  Biochemistry
University of Colorado
Boulder, Colorado

Barbara J. Brandhuber
Department of Chemistry and
  Biochemistry
University of Colorado
Boulder, Colorado

Tanya G. Falbel
Department of Molecular, Cellular and
  Developmental Biology
University of Colorado
Boulder, Colorado

David J. FitzGerald, Ph.D.
Microbiologist
Laboratory of Molecular Biology
National Cancer Institute
National Institutes of Health
Bethesda, Maryland

Arthur M. Friedlander, M. D.
Chief
Pathobiology Division
U.S. Army Medical Research Institute of
  Infectious Diseases
Frederick, Maryland

Jill Marie Manske, Ph.D.
Postdoctoral Fellow
Section on Experimental
  Cancer Immunology
Department of Therapeutic
  Radiology
University of Minnesota
Minneapolis, Minnesota

David B. McKay, Ph.D.
Associate Professor
Department of Chemistry and
  Biochemistry
University of Colorado
Boulder, Colorado

Randal E. Morris, Ph.D.
Associate Professor
Department of Anatomy and Cell Biology
University of Cincinnati College of
  Medicine
Cincinnati, Ohio

Sjur Olsnes, M.D.
Senior Scientist
Department of Biochemistry
Institute for Cancer Research
Montebello, Oslo, Norway

Ira Pastan, M.D.
Chief
Laboratory of Molecular Biology
National Cancer Institute
National Institutes of Health
Bethesda, Maryland

Ole William Petersen, M. D.
Assistant Professor
Structural Cell Biology Unit
Department of Anatomy
University of Copenhagen
Copenhagen, Denmark

Catharine B. Saelinger, Ph.D.
Professor
Department of Molecular Genetics,
  Biochemistry, and Microbiology
University of Cincinnati
Cincinnati, Ohio

Kirsten Sandvig, Ph.D.
Senior Scientist
Department of Biochemistry
Institute for Cancer Research
Norwegian Radium Hospital
Oslo, Norway

Lance L. Simpson, Ph.D.
Professor and Chief
Division of Environmental Medicine and
  Toxicology
Jefferson Medical College
Philadelphia, Pennsylvania

Daniel A. Vallera, Ph.D.
Professor and Director
Section on Experimental Cancer
  Immunology
Department of Therapeutic Radiology
University of Minnesota
Minneapolis, Minnesota

Bo van Deurs, Ph.D.
Associate Professor
Structural Cell Biology Unit
Department of Anatomy
Panum Institute
Copenhagen, Denmark

TABLE OF CONTENTS

Chapter 1

TOXIN STRUCTURE AND FUNCTION

Catharine B. Saelinger

TABLE OF CONTENTS

I. INTRODUCTION

Bacterial toxins have long been recognized as virulence factors produced by different pathogenic bacteria. Poisonous toxins also are produced by certain plants. Bacterial toxins are responsible for diseases that affect many millions of people each year, especially in Third World countries. In most instances, these toxins are very potent and require only a relatively small number of molecules to affect a cell. Initial studies on these toxic proteins are directed at determining their mechanism of action and their effect in an animal model or on cells growing in culture. Genetic manipulations have allowed us to more carefully dissect structure:function relationships. In many instances, the steps involved in the binding, internalization, and subsequent processing of a toxin by a mammalian cell are being studied. Intracellular trafficking of protein toxins is the theme of this book.

The chapters in this book deal with the trafficking of several bacterial and plant toxins in mammalian cells. The level of knowledge available for each toxin differs, but it is seen that they all interact with the eukaryotic cell surface, and are thought to be transported across a cell membrane to exert an effect. The question can be raised as to why a cell has evolved a receptor for a protein which ultimately damages it. Indeed, it is most likely that these toxins enter mammalian cells by exploiting the receptors and entry mechanisms which exist for another purpose, i.e., these toxic proteins are "opportunistic" agents which have usurped the normal internalization machinery of the cell. Thus, to understand entry and trafficking of toxins, one must be familiar with the intracellular transport of physiologically relevant ligands by mammalian cells.

II. PATHWAYS IN RECEPTOR-MEDIATED ENDOCYTOSIS

Two types of endocytosis are generally recognized in mammalian cells, namely, phagocytosis and pinocytosis.[1] Phagocytosis is characterized by the uptake of large particles, such as bacteria or dead cells and is usually associated with neutrophils and macrophages. In contrast, pinocytosis describes the formation of smaller vesicles with the subsequent internalization of extracellular fluid and solutes dissolved therein (fluid phase pinocytosis) as well as any macromolecules which are bound to the membrane at the site where the vesicle is formed (adsorptive endocytosis). Both processes require internalization of extensive amounts of membrane and there is now a great deal of evidence that membrane is rapidly recycled during endocytosis.[1,2] In both phagocytosis and pinocytosis, the final destination is usually the lysosome where the materials which have been internalized are degraded. The large amount of recycling of membrane material seen in endocytosis implies that sorting of membrane material and of vacuolar contents must occur during endocytosis. Many questions remain to be answered about the continuous trafficking of membranes within the mammalian cell, and perhaps protein toxins can be used to address some of these points. The toxins to be examined in detail are believed to enter cells by adsorptive endocytosis and the following discussion will center on this process.

In all typical eukaryotic cells, there is an extensive complex of membrane-bound organelles which belong to the endocytic pathway. This pathway consists of a series of compartments through which materials pass in an orderly and sequential fashion by means of specific transport vesicles and membrane fission and fusion reactions. As already stated, sorting of soluble and membrane-bound molecules occurs in this inward directed pathway. The organelles involved in the endocytic pathway include clathrin coated pits, coated vesicles, endosomes, Golgi-associated vesicles, and lysosomes. Most of these organelles have membrane-associated H^+-ATPases that are responsible for generating an internal acidic environment with the lumen of the organelle.[3,4] Material traversing the endocytic pathway encounters progressively decreasing pH environments as it moves through prelysosomal compartments to lysosomes.

Coated pits and coated vesicles represent the major entry point for most membrane-bound ligands and receptors. The coated pits involved in endocytosis appear as thickened indentations which are covered with protein. Clathrin is the principal component of this coat.[5] While the signals for directing ligand or receptor to clathrin-coated areas are not known, it is felt that clathrin-coated pits regulate receptor-mediated endocytosis. The next station on the endocytic pathway is the endosome. These prelysomal organelles consist of a series of morphologically indistinct vesicular, tubular and multivesicular elements, distributed in the peripheral and perinuclear cytoplasm. Several different types of endosomes have been identified based on buoyant density.[4,6] To date, no specific marker enzyme has been associated with endosomes, although there is some evidence that degradation of protein may occur in endosomes. Endosomes are acidic, with acidification being accomplished by an ATP-driven proton pump.[6] The pH of endosomes ranges from 5 to 6, with the first endosomes identified in the endocytic pathway apparently being less acidic than ones further along the pathway. As will be discussed below, the acidification appears to facilitate the dissociation of receptor and ligand which may occur in endosomes. Vesicles in the Golgi region of the cell also may be part of the endocytic pathway for some ligands. Certain elements of the Golgi complex have proton pumps and are mildly acidic. Lysosomes, the cell's digestive apparatus, represent the final step in some pathways. Material which is delivered to lysosomes is usually degraded. Lysosomes have a pH of 4.5 to 5.5, and they are the most acidic organelle in a mammalian cell. Due to this extreme acidic environment, the integral lysosomal membrane proteins are unique when compared to other membrane proteins, and lysosomal enzymes are active at low pH.[7] In general terms, movement along the endocytic pathway is rapid and involves internalization of large amounts of membrane, with fusion being restricted to a relatively small subpopulation of intracellular organelles.

It should be mentioned that the exocytic pathway involved in the biosynthesis, post-translational modification, and transport of lipids and proteins to the plasma membrane, lysosomes, and secretory granules also is a vacuolar system.[8,9] Here the main organelles are the rough endoplasmic reticulum, Golgi apparatus, secretory vesicles, and other membrane-bound vesicles involved in transport from the Golgi to the cell surface or to intracellular organelles. Portions of the trans-Golgi apparatus are mildly acidic. In addition, there are a variety of coated and noncoated transport vesicles, some of which may be acidic, which function as carriers between organelles.[6] The endocytic and exocytic pathways are not totally distinct, one from another; rather, the two pathways may overlap at the level of the Golgi or the endosome.

Cell biologists have made great advances during the last decade in studying the routing of physiologically relevant ligands utilized by mammalian cells. It has been shown that many serum macromolecules are bound to the cell surface by sepcific receptors and are internalized by receptor-mediated endocytosis. Several distinct intracellular transport pathways have been described depending on the final destination of the ligand (Figure 1). The initial steps in all of the pathways appear to be the same: binding of ligand to receptor on the cell membrane, clustering of ligand-receptor complex to clathrin-coated pits on the cell surface, entry of the complex into coated vesicles, and delivery of the complex to uncoated membrane-bound vesicles, or endosomes. The endosome is now considered the central sorting station of the endocytic pathway, and ligand or receptor may follow one of several routes from this point.[10]

In the first pathway illustrated in Figure 1, ligand and receptor dissociate due to the slightly acidic pH found in the endosomes. From there, ligands, such as low density lipoprotein,[10] asialoglycoproteins,[11] and insulin[12,13] are transported to lysosomes where they are degraded, while the receptors for these ligands recycle to the cell surface to be reused. In a second pathway, originally described for transferrin, both receptor and ligand recycle.[11] Iron-loaded transferrin-receptor complex is internalized; in the acidic environment of the

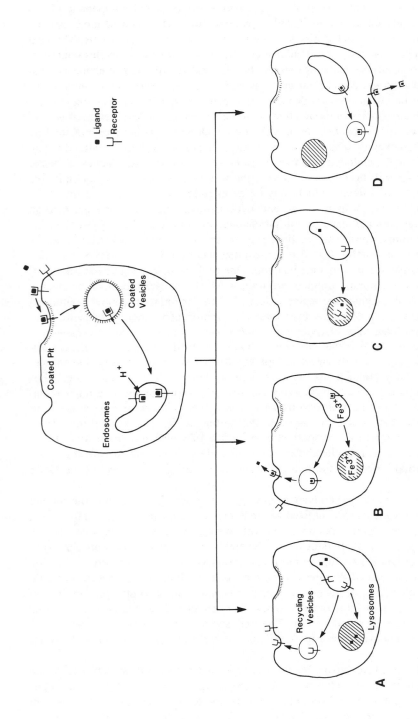

FIGURE 1.   Postulated pathways of receptor-mediated endocytosis in mammalian cells. Similar initial steps are described for all pathways, i.e., binding of ligand to receptor, movement to coated pits, entry of receptor-ligand complex into coated vesicles, and delivery to endosomes. From this point, the ligand-receptor complex can take any of the routes described.

endosome, the iron dissociates from transferrin, but the apotransferrin-receptor complex itself does not dissociate; instead it returns to the cell surface, where, at a neutral pH, the apotransferrin dissociates to be reused by the cells. Transferrin receptor also can be internalized in the absence of ligand; it is thought that some of this receptor traverses the Golgi region of the cell, where it is resialylated.[14,15]

In a third pathway, both receptor and ligand are degraded. This pathway has been best described for epidermal growth factor. Here the receptor-ligand complex reaches the endosomes and presumably dissociates;[16] however, in some cells, receptor and ligand are cotransported to lysosomes where both are degraded.[17,18] Other ligands, such as polymeric IgA are taken up by polarized cells at the baso-lateral surface, are transported across the cell, bound to their receptor, and are released at the apical surface. This last pathway is often referred to as transcytosis. The fact that there are at least four distinct endocytic routes for transporting ligand and receptor suggest that cells have mechanisms to sort molecules after internalization. The molecular mechanisms which dictate dissociation of one type of receptor-ligand complex in endosomes, while allowing another type to remain associated, are unknown. Equally obscure are the cellular mechanisms that shuttle receptor to lysosomes or direct it to recycle to the cell surface. Examination of a variety of receptors has not pointed to a specific feature which is responsible for directing movement, although the oligomeric state of the receptor appears to be able to modulate intracellular movement.[19] In addition, there is evidence that there is a discrete recognition feature, e.g., a tyrosine, in the cytoplasmic domains of certain cell surface receptors which are internalized through coated pits.[20] The nature of this recognition feature is unknown. Further understanding of the intracellular trafficking of toxins and of their receptors should help to elucidate the pathway taken by physiologically essential ligands in mammalian cells.

Several viruses also have been shown to enter mammalian cells and be transported along the pathways described for physiologically essential ligands. Enveloped viruses, such as Vesicular Stomatitis Virus and Semliki Forest Virus bind to the cell surface and are internalized. The acidic environment of the endosome induces a conformational change in the viral fusion protein which initiates viral penetration, and the nucleocapsid is released into the cytosol.[21] Productive infection with certain picornaviruses also appears to require exposure to an acidic pH. Polio viruses are taken up by receptor-mediated endocytosis from coated pits; in a low pH environment, the hydrophobic domains of the viral capsid are exposed, and infectious RNA is transferred to the cytosol.[22-24]

In addition to entry via clathrin-coated pits, it is also proposed that some ligands internalized by adsorptive endocytosis bind to cells and are internalized through noncoated invaginations of the plasma membrane. This was initially described for tetanus toxin-gold, cholera toxin-gold, and anti-HLA antibody, and is described in detail for ricin (Chapter 6). While the intracellular pathway for ligands which enter cells through coated pits has been extensively studied, the routes taken by ligands, which are believed to enter through noncoated areas, are just beginning to be examined in detail. Evidence suggests that ligands entering mammalian cells by either route eventually are found in similar intracellular organelles.

III. REVIEW OF TOXIN STRUCTURE AND FUNCTION

The toxins considered here all have evolved a similar overall structure and are made up of two components: a binding component which is responsible for binding the protein to the cell surface and an active component which is responsible for the pharmacological effect of the toxin. In some, but not all, cases, the active component has been shown to be an enzyme. In most cases, the two functions are encoded in different polypeptide subunits. The active subunit is usually composed of one polypeptide chain, and the binding subunit,

TABLE 1
Comparison of Toxins of Bacterial and Plant Origin

Toxin	M_r (kDa)	Structure	Target
Alter Protein Synthesis			
1. Diphtheria	58	A,B	EF2
2. Pseudomonas exotoxin A	66.5	A,B	EF2
3. Ricin, abrin	62	A,B	60S ribosome
Clostridial			
1. Botulinum neurotoxins	150	L,H.H$_2$	Not determined
2. Tetanus	150	L,H,H$_2$	Not determined
3. Botulinum C$_2$	150	Binary (I + II)	Actin
Stimulate Adenylate Cyclase			
1. Cholera	82	A$_1$,A$_2$,5B	G$_s$ (α subunit)[a]
2. *E. coli* heat labile	91	A$_1$,A$_2$,5B	G$_s$ (α subunit)[a]
3. Pertussis	117	A,B	G$_i$ (α subunit)[a]
4. Anthrax	85	EF + PA	Not determined

[a]   Adenylate cyclase complex.

of one to six chains. In addition to binding and enzyme activity, the ability to traverse a membrane may be associated with one of the two subunits. As is obvious from Table 1, there is variety in the way the whole toxins are formed. The structures and mechanisms of action of these toxins have been described in a number of reviews;[29-31] therefore, we present only an overview of the toxins.

A. TOXINS WHICH BLOCK PROTEIN SYNTHESIS
1. Plant Toxins: Abrin, Ricin, and Modeccin

Why plants produce toxins is not known, although they may have a role in protecting the plants or their seedlings agains animals. The toxins are located in different parts of the producing plants. For example, modeccin is found in the roots, and abrin and ricin are present in seeds. Plant toxins are among the most poisonous compounds known. They are toxic for most eukaryotic cells studied and cause serious intoxication following oral or parenteral administration.[29]

The plant toxins discussed here are similar in structure. All are glycoproteins and are composed of two polypeptide chains linked by a disulfide bond. The A-chain has the enzymatic activity, and the B-chain has lectin-like properties and binds to carbohydrates which have a terminal galactose. In all cases, the A-chains block protein synthesis by interacting with the 60S ribosomal subunit.[29,32] This action requires neither cofactors nor energy. Recently Endo et al.[33] have reported that ricin, abrin, and modeccin act directly on ribosomal RNA, not ribosomal proteins. They show that ricin A-chain is an RNA N-glycosidase which inactivates ribosomes by modifying a single nucleoside residue (A_{4324}) in 28S rRNA.[34] While the amino acid residues essential for A-chain activity have not been found, it has been reported that arginine residue(s) are involved in the inhibition of protein synthesis.[35]

The amino acid sequence and carbohydrate composition of both ricin A- and B-chains have been determined. Both chains are coded by a single messenger RNA. The nucleotide sequence for both the cDNA and the genomic DNA has been obtained.[36,37] Both nucleotide sequences show ricin is synthesized as a proenzyme with a leader sequence and a 12-amino acid peptide which links the A- and B-chain. The leader sequence and linking peptide are removed during processing to the mature form. The three-dimensional structure of ricin has been determined by X-ray crystalography at the 2.8 Å resolution.[38] The enzymatic A-chain

NAD + ELONGATION FACTOR 2 $\rightleftharpoons$ ADP RIBOSE~ELONGATION
FACTOR 2 + NICOTINAMIDE

FIGURE 2.   Mechanism of action of pseudomonas exotoxin A and diphtheria toxin.

is a globular protein, while the B-chain lectin is composed of two separate folding domains (each of which bind one lactose disaccharide). Ricin is isolated as a dimer consisting of two glycoproteins (M_r = 30,625 and 31,431) linked by a disulfide bond. Recent studies on ricin suggest that the strong affinity of A- and B-chains for each other is mediated by hydrophobic forces and that association of the two chains is required for toxicity. The disulfide bond between the A- and B-chains does not appear critical for toxicity.[39] Besides the role in binding, the B-chain is thought to facilitate the entrance of the A-chain into the cell cytoplasm.[40] In addition, a hydrophobic region on the ricin A-chain also may be involved in entry of this chain into the cytosol.[41]

2. Diphtheria Toxin and Pseudomonas Exotoxin A

The toxin produced by *Corynebacterium diphtheriae* is responsible for the majority of the symptoms associated with the disease diphtheria. Mass immunization with diphtheria toxoid (inactivated toxin) leads to antitoxin formation and has resulted in the virtual eradication of the disease in developed countries. Diphtheria toxin has been studied extensively. It is synthesized as a single polypeptide chain (M_r = 58,342), which can be proteolytically cleaved in an arginine-rich region into the A fragment (M_r = 21,145) and the B fragment (M = 37,240); these fragments remain linked by a single disulfide bridge.[30,42] Upon reduction of the disulfide bond, the two fragments dissociate. Fragment B is involved in binding diphtheria toxin to receptors on the cell surface. There are two sites on the B fragment, a polyphosphate binding site (P site) and a second site (X site) believed to be involved in receptor binding.[43] The binding fragment also has a hydrophobic region which is important for entry of the A fragment. The A fragment has enzyme activity (see below). No enzyme activity has been associated with fragment B, and neither fragment alone is toxic for cells in culture or for animals.[30]

Pseudomonas exotoxin A is one of several extracellular products produced by *Pseudomonas aeruginosa* that may contribute to the disease process. Its contribution depends upon the type of infection. Evidence for toxin production in human infections comes from the work of Pollack et al.,[44,45] who found high titers of hemagglutinating and neutralizing antibodies in sera of patients who had recovered from serious pseudomonas infections, and low antitoxin titers in several patients with fatal infections.

Pseudomonas exotoxin is produced as a single chain polypeptide (M_r = 66,583), and must undergo a conformational change to express enzymatic activity.[46] Recent genetic studies have identified an enzymatic domain and a binding domain and a possible translocation domain.[47,48] The structures of diphtheria and pseudomonas toxin are examined in detail in Chapter 2.

Pseudomonas exotoxin A and diphtheria toxin act identically on the molecular level. Both toxins are ADP-ribosyl transferase enzymes and ADP-ribosylate elongation factor 2. Elongation factor 2 is a mammalian cell enzyme involved in protein synthesis; specifically it translocates the new petidyl-tRNA from the A site to the P site on the ribosome, as the ribosome moves along the messenger RNA molecule. When elongation factor 2 is inactivated, protein synthesis stops. Diphtheria toxin and pseudomonas exotoxin ADP-ribosylate a unique amino acid, diphthamide, which has been found only in elongation factor 2.[49] Specifically both toxins catalyze the cleavage of NAD+ to nicotinamide and ADP-ribose, and the subsequent covalent linkage of ADP-ribose to diphthamide (Figure 2). Cells which lack diphthamide are resistant to both toxins.[50]

Recently Carroll and Collier[51] have shown amino acid sequence homology between the enzymatic domains of pseudomonas and diphtheria toxins. In both instances, a glutamic acid has been shown to function as the active site residue; in the case of pseudomonas toxin, it is glu 553,[52] and glu 148 for diphtheria toxin. These glutamic acid residues are in the NAD binding site, and conversion of the glutamate to aspartic acid, using site-directed mutagenesis results in over a 100-fold reduction in enzyme activity.[53] Of interest, a glutamic acid residue also has been shown to be critical for the activity of Shiga-like toxin,[54] a bacterial toxin belonging to another class of ADP-ribosylating toxins.

B. NEUROTOXINS: TETANUS AND BOTULINUM

Both botulinum and tetanus toxins are produced by anaerobic spore-forming bacilli, *Clostridium botulinum* and *C. tetani,* respectively. Tetanus toxin is synthesized by *C. tetani* growing in a wound site. The toxin is taken up at neuromuscular junctions and is then transported by retrograde axonal transport to synapses. There it acts presynaptically to block the release of inhibitory transmitters which results in the characteristic paralysis associated with tetanus. Botulinum toxin in contrast is usually ingested preformed in contaminated food. Alternatively, in the case of infant botulism, the toxin is produced by organisms which colonize the infant's intestinal tract. Botulinum neurotoxins block transmitter release from peripheral cholinergic nerve endings, leading to muscular weakness and paralysis.[31,55] Tetanus and botulinum neurotoxins have similar structural features and may have similar subcellular modes of action.

There are seven botulinum neurotoxins (A,B,C_1,D,E,F, and G) which act at the neuromuscular junction to block acetylcholine release. An eighth toxin (C_2) which is unique in its pharmacologic action and its structure also has been identified and is called botulinum binary toxin.[56] *C. botulinum* neurotoxins are synthesized as single polypeptide chains (M_r ~150 kDa). Proteolytic cleavage yields a dichain molecule, composed of a light chain (M_r ~50 kDa) and a heavy chain (M_r ~100 kDa) linked by a disulfide bond. The single chain molecule has minimal activity, while the dichain molecule is fully active. Reduction of the disulfide bond results in loss of toxicity.[57] Upon further cleavage of the nicked neurotoxin molecule with papain, the H chain is split into two fragments. The carboxy terminus is now free and the amino terminus remains covalently linked to the light chain. It is believed that the light chain is responsible for stopping nerve cell function (the ''poisoning domain'') while the carboxyterminus of the heavy chain mediates binding to the target cell, and the amino terminus of the heavy chain is involved in channel formation. Analogous to other toxins, no chain by itself has neurotoxic activity.

The botulinum neurotoxins have been thought to act enzymatically based on their structure, extreme potency, and long-lasting action. Only recently, however, has any enzymatic activity been associated with this group of toxins. Ohashi and colleagues[58,59] and Matsuoka et al.[60] have shown that type C_1 and D neurotoxins may ADP-ribosylate a 21-kDa protein in membranes obtained from mouse brain. No ADP ribosyl-transferase activity is detectable for type A, B or E toxins. At present, it is not known if such activity is masked or is lacking in these toxins.

Tetanus toxin acts predominantly on central inhibitory synapses, although it too can block neuromuscular transmission. The synthesis and proteolytic processing of tetanus toxin is similar to that of the botulinum neurotoxins.[61] Tetanus toxin is synthesized as an inactive single-chain polypeptide having a molecular weight of ~150 kDa and is cleaved by endogenous proteases to yield light and heavy subchains (M_r ~ 50 kDa and 100 kDa). The light and heavy chains are linked by a disulfide bridge. As with the botulinum neurotoxins, the light chain is assumed to be the poisoning domain, while the carboxy terminus of the heavy chain is the binding domain, and the amino terminus of the heavy chain is the translocation domain. The individual chains have no biological activity. No enzymatic activity has yet been ascribed to tetanus toxin.

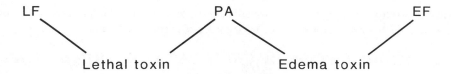

FIGURE 3. Complex structure of edema toxin and lethal toxin. EF is edema factor, LF is lethal factor, and PA is protective antigen.

The structure of the binary toxin produced by *C. botulinum* is different from that of the clostridial neurotoxins. The binary toxin is composed of two dissimilar polypeptides (M_r 50 kDa and 100 kDa), having no interchain covalent bonds.[62] The toxin has a variety of biological activites including induction of vascular permeability, enterotoxic, cytotoxic, and lethal activities. All of these activities require the cooperation of both toxin components. As with the neurotoxins, the heavy chain (component II) has a role in cell surface binding, and the light chain (component I) has the ability to enzymatically ADP-ribosylate nonmuscle actin.[63-65]

C. ANTHRAX TOXIN

Virulent strains of *Bacillus anthracis* produce two virulence factors: a poly-D-glutamic acid capsule and a tripartite toxin which is composed of proteins designated edema factor (EF), lethal factor (LF), and protective antigen (PA). These combine to form two toxic complexes, a lethal toxin and an edema toxin (Figure 3). The lethal toxin is composed of PA or factor II and LF or factor III. This toxin is lethal for several animal species including rats and mice. Neither LF nor PA alone is toxic; however, a combination of the two causes severe pulmonary edema and death of rats within 40 to 60 min. The edema toxin is composed of PA and EF (factor I). Edema toxin produces localized edema in the skin of guinea pigs and rabbits. It is now known to be a calmodulin-dependent adenylate cyclase.[66,67] The combination of edema factor plus protective antigen can suppress polymorphonuclear leukocyte function and impair host resistance, resulting in an increased susceptibility to infection with *B. anthracis*.[68] PA is believed to act as a "receptor" for EF and LF. PA is considered an antigen which elicits protection against anthrax infection in animals immunized with it. The PA gene has been cloned into *Escherichia coli*[69] and *B. subtilis*.[70]

D. CONJUGATE TOXINS

Immunotoxins are made by chemically linking antibodies to an intact toxin molecule or to the A-chain of a toxin, and thus forming a hybrid molecule which has the specificity of the immunological ligand and the toxicity of the toxin. The antibody component directs the hybrid to the target cell; once the hybrid molecule is bound to the cell surface, the toxic moiety may enter the cell and stop protein synthesis. Alternate forms of conjugate toxins have been made, using a physiologically relevant ligand, such as transferrin, to direct the conjugate movement. Ricin A-chain, diphtheria toxin (fragment A or intact), and pseudomonas exotoxin A have been used most frequently in the construction of conjugates.[71-73]

The linkage between the toxin and the targeting moiety influences the effectiveness of the conjugates. Most frequently a thiol-containing heterobifunctional reagent such as *N*-succinimidyl-3-(2-pyridyldithio) propionate (SPDP) is used as cross-linking reagent. SPDP reacts with free amino groups in the antibody and by disulfide exchange can attach either the SH-containing A chain or the SPDP-derivatized toxin to the antibody. As would be expected, the type of linkage between the antibody and the A-chain or toxin is important in determining toxicity. The bond must be stable enough to maintain the conjugate until it reaches its target cell and is properly internalized. Also, it must be readily broken in an endosome or lysosome; if not, the toxicity of the conjugate will be abolished. Formation and routing of toxic conjugates in mammalian cells, and factors which influence the routing of the conjugates or the expression of toxicity are discussed in Chapters 8 and 9.

IV. MODIFICATION OF CELLULAR FUNCTION BY ADP-RIBOSYLATION

Modification of proteins by ADP-ribosylation is known to occur in a variety of situations in nature. The target of the ADP-ribosyltransferase in eukaryotic cells can be within the cytoplasm or in the nucleus. The ADP-ribosylating toxins to be discussed here all have cytoplasmic substrates for their action. As already described, diphtheria and pseudomonas toxins modify elongation factor 2 resulting in inhibition of protein synthesis.[30] ADP-ribosylation by several bacterial toxins affects adenylate cyclase regulation. Cholera toxin, produced by *Vibrio cholerae,* and a heat-labile toxin produced by *Escherichia coli,* stimulate adenylate cyclase by ADP-ribosylating the stimulatory guanyl nucleotide-binding regulatory subunit, G_s. ADP ribosylation of $Gs\alpha$ reduces its affinity for $Gi\beta$ (a subunit of the inhibitory guanine nucleotide binding protein) so that adenylate cyclase activity is constantly expressed.[74] In contrast, pertussis toxin, produced by *Bordetella pertussis,* acts on the inhibitory arm of adenylate cyclase by ADP-ribosylating the inhibitory guanyl nucleotide-binding regulatory protein, G_i.[75] Botulinum C_2 toxin, a nonneurotoxin ADP-ribosylates nonmuscle and smooth muscle actin but not skeletal muscle actin C.[31,63] This occurs in intact cells and results in rounding up of the cells. Lastly, an endogenous mono(ADP-ribosyl)transferase has been found in polyoma virus transformed baby hamster kidney cells and beef liver which is able to transfer ADP-ribose from NAD to the diphthamide residue of elongation factor 2, suggesting a role in normal cellular regulation.[76] This overview suggests that the ability of microbial toxins to alter cellular function in target cells by the ADP-ribosylation of regulatory proteins is widespread and is found in a diverse array of situations. These toxins afford the cell biologist precise tools to study the functions of ADP-ribosylation and of G proteins in mammalian cell regulation.

V. CONCLUSIONS

The toxins discussed in this book all are hypothesized to enter mammalian cells by a receptor-mediated route. Very strong evidence exists for toxins, such as diphtheria toxin, ricin and pseudomonas exotoxin A; in other cases, the evidence is circumstantial. It is already known that the receptor for these toxins and the subsequent route taken by the proteins differs. Ricin, at least in part, enters cells through noncoated invaginations and is shuttled to the Golgi region of the cell. On the other hand, pseudomonas exotoxin is known to enter sensitive cells via coated areas, and move to the Golgi and lysosomes. What determines the routing of toxins, i.e., the receptor itself or the toxin molecule, is not known. Indeed, very little is known about receptors for most toxins, although research is ongoing in that area. Isolation and characterization of these receptors also may help us to determine if the toxins do indeed usurp physiologically important receptors for their own use.

Another question which still remains to be ansered is how the toxin molecules or their active fragments cross the membrane surrounding the compartments involved in internalization. Genetic engineering of toxin molecules allows us to determine the part of the molecule involved in translocation; however, the organelles from which toxin enters the cytosol, the function of pH in this translocation, and the molecular mechanisms involved are far from understood.

In conclusion, studies on the trafficking of toxic proteins in mammalian cells provide a strong bridge between the discipline of cell biology and microbiology. They move far beyond the goals of elucidating toxin action and help us understand how proteins are handled on their journey through the mammalian cell.

REFERENCES

1. **Steinman, R. M., Mellman, I., Muller, W. A., and Cohn, Z. A.,** Endocytosis and the recycling of plasma membrane, *J. Cell Biol.,* 96, 1, 1983.
2. **Mellman, I., Howe, C., and Helenius, A.,** The control of membrane traffic on the endocytic pathway, *Curr. Topics Membr. Res.,* 29, 255, 1987.
3. **Al-Awqati, Q.,** Proton-translocating ATPases, *Annu. Rev. Cell Biol.,* 2, 179, 1986.
4. **Mellman, I., Fuchs, R., and Helenius, A.,** Acidification of the endocytic and exocytic pathways, *Annu. Rev. Biochem.,* 55, 663, 1986.
5. **Pearse, B. M. F.,** The EMBO Medal Review, Clathrin and coated vesicles, *EMBO J.,* 6, 2507, 1987.
6. **Anderson, R. G. W. and Orci, L.,** A view of acidic intracellular compartments, *J. Cell Biol.,* 106, 539, 1988.
7. **von Figura, K. and Hasilik, A.,** Lysosomal enzymes and their receptors, *Annu. Rev. Biochem.,* 55, 167, 1986.
8. **Griffiths, G. and Simons, K.,** The *trans* Golgi network: sorting at the exit site of the Golgi complex, *Science,* 234, 438, 1986.
9. **Dunphy, W. G. and Rothman, J. E.,** Compartmental organization of the Golgi stack, *Cell,* 42, 13, 1985.
10. **Goldstein, J. L., Brown, M. S., Anderson, R. G. W., Russell, D. W., and Schneider, W. J.,** Receptor-mediated endocytosis: concepts emerging from the LDL receptor system, *Annu. Rev. Cell Biol.,* 1, 1, 1985.
11. **Ciechanover, A., Schwartz, A. L., and Lodish, H. F.,** Sorting and recycling of cell surface receptors and endocytosed ligands: the asialoglycoprotein and transferrin receptors, *J. Cell Biochem.,* 23, 107, 1983.
12. **Czech, M. P.,** The nature and regulation of the insulin receptor: structure and function, *Annu. Rev. Physiol.,* 47, 357, 1985.
13. **Bergeron, J. J. M., Cruz, J., Khan, M. N., and Posner, B. I.,** Uptake of insulin and other ligands into receptor-rich endocytic components of target cells: The endosomal apparatus, *Annu. Rev. Physiol.,* 47, 383, 1985.
14. **Stein, B. S. and Sussman, H. H.,** Demonstration of two distinct transferrin receptor recycling pathways and transferrin-independent receptor internalization in K562 cells, *J. Biol. Chem.,* 261, 10319, 1986.
15. **Woods, J. W., Doriaux, M., and Farquhar, M. G.,** Transferrin receptors recycle to *cis* and middle as well as *trans* Golgi cisternae in Ig-secreting myeloma cells, *J. Cell Biol.,* 103, 277, 1986.
16. **DiPaola, M. and Maxfield, F. R.,** Conformational changes in the receptors for epidermal growth factor and asialoglycoproteins induced by the mildly acidic pH found in endocytic vesicles, *J. Biol. Chem.,* 259, 9163, 1984.
17. **Carpenter, G.,** Receptors for epidermal growth factor and other polypeptide mitogens, *Annu. Rev. Biochem.,* 56, 881, 1987.
18. **Carpenter, G. and Zendegui, J. G.,** Epidermal growth factor, its receptor, and related proteins, *Exp. Cell Res.,* 164, 1, 1986.
19. **Matlin, K. S., Sibbens, J., and McNeil, P. L.,** Reduced extracellular pH reversibly inhibits oligomerization, intracellular transport, and processing of the influenza hemagglutinin in infected Madin-Darby Canine Kidney cells, *J. Biol. Chem.,* 263, 11478, 1988.
20. **Lazarovits, J. and Roth, M.,** A single amino acid change in the cytoplasmic domain allows the influenza virus hemagglutinin to be endocytosed through coated pits, *Cell,* 53, 743, 1988.
21. **Kiellian, M. and Helenius, A.,** Entry of alpha-viruses, in *The Virus Series,* Fraenkel-Conrad, H. and Wagner, R., Eds., Plenum Press, New York, 1985.
22. **Madshus, I. H., Olsnes, S., and Sandvig, K.,** Mechanism of entry into the cytosol of poliovirus type 1: requirement for low pH, *J. Cell Biol.,* 98, 1194, 1984.
23. **Madshus, I. H., Olsnes, S., and Sandvig, K.,** Requirements for entry of poliovirus RNA into cells at low pH, *EMBO J.,* 3, 1945, 1984.
24. **Madshus, I. H., Sanvig, K., Olsnes, S., and van Deurs, B.,** Effect of reduced endocytosis induced by hypotonic shock and potassium depletion on the infection of Hep2 cells by picornaviruses, *J. Cell. Physiol.,* 131, 14, 1987.
25. **Huet, C., Ash, J. F., and Singer, S. J.,** The antibody-induced clustering and endocytosis of HLA antigens on cultured human fibroblasts, *Cell,* 21, 429, 1980.
26. **Montesano, R., Roth, J., Robert, A., and Orci, L.,** Non-coated membrane invaginations are involved in binding and internalization of cholera and tetanus toxins, *Nature,* 296, 651, 1982.
27. **Tran, D., Carpentier, J.-L., Sawano, F., Gorden, P., and Orci, L.,** Ligands internalized through coated or noncoated invaginations follow a common intracellular pathway, *Proc. Natl. Acad. Sci. U.S.A.,* 84, 7957, 1987.
28. **Brown, D., Weyer, P., and Orci, L.,** Nonclathrin-coated vesicles are involved in endocytosis in kidney collecting duct intercalated cells, *Anat. Rec.,* 218, 237, 1987.
29. **Olsnes, S. and Pihl, A.,** Toxic lectins and related proteins, in *Molecular Action of Toxins and Viruses,* Cohen and van Heyningen, Eds., Elsevier, Amsterdam, 1982.

30. **Middlebrook, J. L. and Dorland, R. B.,** Bacterial toxins: cellular mechanisms of action, *Microbiol. Rev.,* 48, 199, 1984.

31. **Simpson, L. L.,** Molecular pharmacology of botulinum toxin and tetanus toxin, *Annu. Rev. Pharmacol. Toxicol.,* 26, 427, 1986.

32. **Olsnes, S. and Sandvig, K.,** Entry of polypeptide toxins into animal cells, in *Endocytosis,* Pastan, I. and Willingham, M. C., Eds., Plenum Press, New York, 1985.

33. **Endo, Y., Mitsui, K., Motizuki, M., and Tsurugi, K.,** The mechanism of action of ricin and related toxic lectins on eukaryotic ribosomes, *J. Biol. Chem.,* 262, 5908, 1987.

34. **Endo, Y. and Tsurugi, K.,** RNA *N*-glucosidase activity of ricin A-chain: mechanism of action of the toxic lectin on eukaryotic ribosomes, *J. Biol. Chem.,* 262, 8128, 1987.

35. **Watanabe, K. and Funatsu, G.,** Involvement of arginine residues in inhibition of protein synthesis by ricin A-chain, *FEBS Lett.,* 204, 219, 1986.

36. **Lamb, F., Roberts, L., and Lord, J. M.,** Nucleotide sequence of cloned cDNA coding for preproricin, *Eur. J. Biochem.,* 148, 265, 1985.

37. **Halling, K., Halling, A., Murray, E., Ladin, B., Houston, L., and Weaver, R.,** Genomic cloning and characterization of a ricin gene from *Ricinus communis, Nucleic Acids Res.,* 13, 8019, 1985.

38. **Montfort, W., Villafranca, J. E., Monzingo, A. F., Ernst, S. R., Katzin, B., Rutenber, E., Xuong, N. H., Hamlin, R., and Robertus, J. D.,** The three-dimensional structure of ricin at 2.8 Å, *J. Biol. Chem.,* 262, 5398, 1987.

39. **Lewis, M. S. and Youle, R. J.,** Ricin subunit association: Thermodynmics and the role of the disulfide bond in toxicity, *J. Biol. Chem.,* 261, 11571, 1986.

40. **Simmons, B. M., Stahl, P. D., and Russell, J. H.,** Mannose receptor-mediated uptake of ricin toxin and ricin A chain by macrophages, *J. Biol. Chem.,* 261, 7912, 1986.

41. **Utsumi, T., Aizono, Y., and Funatsu, G.,** Interaction of ricin and its constitutive polypeptides with dipalmitoylphosphatidyl-choline vesicles, *Biochim. Biophys. Acta,* 772, 202, 1984.

42. **Neville, D. M., Jr. and Hudson, T. H.,** Transmembrane transport of diphtheria toxin, related toxins, and colicins, *Annu. Rev. Biochem.,* 55, 195, 1986.

43. **Eidels, L., Ross, L., and Hart, D.,** Diphtheria toxin-receptor interaction: A polyphosphate-insensitive diphtheria toxin-binding domain, *Biochem. Biophys. Res. Commun.,* 109, 493, 1982.

44. **Pollack, M., Callahan, L. T., III, and Taylor, N. S.,** Neutralizing antibody to *Pseudomonas aeruginosa* exotoxin in human sera: evidence for *in vivo* toxin production during infections, *Infect. Immun.,* 14, 942, 1976.

45. **Pollack, M., Taylor, N. S., and Callahan, L. T., III,** Exotoxin production by clinical isolates of *Pseudomonas aeruginosa, Infect. Immun.,* 15, 776, 1977.

46. **Leppla, S. H., Martin, O. C., and Muehl, L. A.,** The exotoxin of *P. aeruginosa:* a proenzyme having an unusual mode of activation, *Biochem. Biophys. Res. Commun.,* 81, 532, 1978.

47. **Guidi-Rontani, C. and Collier, R. J.,** Exotoxin A of *Pseudomonas aeruginosa:* evidence that domain I functions in receptor binding, *Mol. Microbiol.,* 1, 67, 1987.

48. **Hwang, J., Fitzgerald, D. J., Adhya, S., and Pastan, I.,** Functional domains of Pseudomonas exotoxin identified by deletion analysis of the gene expressed in *E. coli, Cell,* 48, 129, 1987.

49. **Van Ness, B.G., Howard, J. B., and Bodley, J. W.,** ADP-ribosylation of elongation factor 2 by diphtheria toxin, *J. Biol. Chem.,* 255, 10710, 1980.

50. **Moehring, T. J. and Moehring, J. M.,** Selection and characterization of cells resistant to diphtheria toxin and pseudomonas exotoxin A: presumptive translational mutants, *Cell,* 11, 447, 1977.

51. **Carroll, S. F. and Collier, R. J.,** Amino acid sequence homology between the enzymic domains of diphtheria toxin and *Pseudomonas aeruginosa* exotoxin A, *Mol. Microbiol.,* 2, 293, 1988.

52. **Douglas, C. M. and Collier, R. J.,** Exotoxin A of *Pseudomona aeruginosa.* Substitution of glutamic acid 553 with aspartic acid drastically reduces toxicity and enzymatic activity, *J. Bacteriol.,* 169, 4967, 1987.

53. **Tweten, R. K., Barbieri, J. T., and Collier, R. J.,** Diphtheria toxin: effect of substituting aspartic acid for glutamic acid 148 on ADP-ribosyltransferase activity, *J. Biol. Chem.,* 260, 10392, 1985.

54. **Hovde, C. J., Calderwood, S. B., Mekalanos, J. J., and Collier, R. J.,** Evidence that glutamic acid 167 is an active-site residue of Shiga-like toxin I, *Proc. Natl. Acad. Sci. U.S.A.,* 85, 2568, 1988.

55. **Sugiyama, H.,** *Clostridium botulinum* neurotoxin, *Microbiol. Rev.,* 44, 419, 1980.

56. **Simpson, L. L.,** A comparison of the pharmacological properties of *Clostridium botulinum* type C_1 and C_2 toxins, *J. Pharmacol. Exp. Ther.,* 223, 695, 1982.

57. **DasGupta, B. R. and Sugiyama, H.,** A common subunit structure in *Clostridium botulinum* type A, B and E toxins, *Biochem. Biophys. Res. Commun.,* 48, 108, 1972.

58. **Ohashi, Y. and Narumiya, S.,** ADP-ribosylation of a M_r 21,000 membrane protein by type D botulinum toxin, *J. Biol. Chem.,* 262, 1430, 1987.

59. **Ohashi, Y., Kamiya, T., Fujiwara, M., and Narumiya, S.,** ADP-ribosylation by type C1 and D botulinum neurotoxins: stimulation by guanine nucleotides and inhibition by guanidino-containing compounds, *Biochem. Biophys. Res. Commun.,* 142, 1032, 1987.

60. **Matsuoka, I., Syuto, B., Kurihara, K., and Kubo, S.**, ADP-ribosylation of specific membrane proteins in pheochromocytoma and primary-cultured brain cells by botulinum neurotoxins type C and D, *FEBS Lett.*, 216, 295, 1987.

61. **Robinson, J. P. and Hash, J. H.**, A review of the molecular structure of tetanus toxin, *Mol. Cell. Biochem.*, 48, 33, 1982.

62. **Takasawa, T., Ohishi, I., and Shiokawa, H.**, Amino-acid composition of components I and II of botulinum C_2 toxin, *FEMS Microbiol. Lett.*, 40, 51, 1987.

63. **Aktories, K., Bärmann, M., Ohishi, I., Tsuyama, S., Jakobs, K., and Habermann, E.**, Botulinum C2 toxin ADP-ribosylates actin, *Nature*, 322, 390, 1986.

64. **Aktories, K., Ankenbauer, T., Schering, B., and Jakobs, K. H.**, ADP-ribosylation of platelet actin by botulinum C2 toxin, *Eur. J. Biochem.*, 161, 155, 1986.

65. **Ohishi, I. and Tsuyama, S.**, ADP-ribosylation of nonmuscle actin with the component I of C_2 toxin, *Biochem. Biophys. Res. Commun.*, 136, 802, 1986.

66. **Leppla, S. H.**, Anthrax toxin edema factor: a bacterial adenylate cyclase that increases cyclic AMP concentrations in eukaryotic cells, *Proc. Natl. Acad. Sci. U.S.A.*, 79, 3162, 1982.

67. **Leppla, S. H., Ivins, B., and Ezzell, J. W., Jr.**, Anthrax toxin, in *Microbiology—1985* Leive, L., Ed., American Society for Microbiology, Washington, D.C., 1985.

68. **O'Brien, J., Friedlander, A., Dreier, T., Ezzell, J., and Leppla, S.**, Effects of anthrax toxin components on human neutrophils, *Infect. Immun.*, 47, 306, 1985.

69. **Vodkin, M. H. and Leppla, S. H.**, Cloning of the protective antigen gene of *Bacillus anthracis*, *Cell*, 34, 693, 1983.

70. **Ivins, B. E. and Welkos, S. L.**, Cloning and expression of the *Bacillus anthracis* protective antigen gene in *Bacillus subtilis*, *Infect. Immun.*, 54, 537, 1986.

71. **Olsnes, S. and Pihl, A.**, Chimeric toxins, *Pharmacol. Ther.*, 15, 355, 1982.

72. **Pastan, I., Willingham, M. C., and FitzGerald, D. J. P.**, Immunotoxins, *Cell*, 47, 641, 1986.

73. **Vitetta, E. S. and Uhr, J. W.**, Immunotoxins, *Annu. Rev. Immunol.*, 3, 197, 1985.

74. **Quinn, T. C., Bender, B. S., and Bartlett, J. G.**, New developments in infectious diarrhea, *Dis. Mon.*, 32, 174, 1986.

75. **Hewlett, E. L.**, Biological effects of pertussis toxin and *Bordetella* adenylate cyclase on intact cells and experimental animals, in *Microbiology, 1984*, Leive, L. and Schlessinger, D., Eds., American Society for Microbiology, Washington, D.C.,

76. **Lee, H. and Iglewski, W. J.**, Cellular ADP-ribosyltransferase with the same mechanisms of action as diphtheria toxin and pseudomonas toxin A, *Proc. Natl. Acad. Sci. U.S.A.*, 81, 2703, 1984.

Chapter 2

STRUCTURES OF PSEUDOMONAS AND DIPHTHERIA TOXINS

David B. McKay, Viloya S. Allured, Barbara J. Brandhuber, and Tanya G. Falbel

TABLE OF CONTENTS

I. INTRODUCTION

Exotoxin A of *Pseudomonas aeruginosa* (pseudomonas toxin) and the major exotoxin of *Corynebacterium diphtheriae* (diphtheria toxin) appear to follow the same overall pathway in inflicting toxicity on a target cell (Figure 1). A proenzyme form of the toxin, as secreted from the bacterium, binds a receptor on a target cell surface. The toxin-receptor complex is then internalized via receptor-mediated endocytosis into endosomal vesicles, where the low pH environment is thought to induce a conformational change in the toxin molecule that results in translocation of at least an enzymatic moiety into the cell cytoplasm.[1,2] Finally, the activated enzyme (fragment?) catalyzes the ADP-ribosylation of a specific modified histidine (named "diphthamide") on protein synthesis elongation factor 2 (EF-2)[3,4] (Figure 2). The toxins thus manifest three relatively distinct biochemical activities: receptor binding, membrane translocation, and enzymatic ADP-ribosylation of EF-2.

The similarity of overall intoxication pathway, and the observation that the kinetic constants for the ADP-ribosyl transferase reaction are identical to within experimental uncertainty for the two toxins[5,6] suggested historically that the proteins might be very similar in primary and tertiary structure. However, a plethora of evidence argues against a strong homology, including (1) the difference in spectrum of cells to which the proteins are toxic[7] — for example, mouse cells are dramatically more sensitive to pseudomonas toxin than to diphtheria toxin; (2) the lack of strong antigenic cross-reactivity;[4] (3) the different overall construction of the two proteins—it has been established for some time that biochemical functions in diphtheria toxin map as: enzymatic domain, membrane translocation domain, and receptor-binding domain proceeding from amino to carboxy terminus,[8] whereas it was recently suggested by three-dimensional structure and confirmed by deletion mapping of the *P. aeruginosa* exotoxin A gene, that in pseudomonas toxin the biochemical functions follow the opposite order: receptor binding, membrane translocation, and enzymatic domains proceeding from amino to carboxy terminus.[9,10] The lack of strong overall homology between the two toxins was demonstrated explicitly when the nucleotide sequences of their genes,[11-13] and by inference their amino acid sequences, were determined; it was reported that there is no significant homology between the two sequences.

A natural question to pose — for which an affirmative response is not an *a priori* certainty — is whether knowing the three-dimensional structure of either of the molecules will suggest mechanisms for its biochemical function. Further, the apparent dissimilarity of the two toxin molecules, at least at the level of primary structure, leads to the intriguing question of to what extent they have similar mechanisms of action, and whether knowing and comparing the structures of both can yield insights beyond those elicited from a single structure.

Both molecules consist of relatively large single polypeptide chains, of length 535 amino acids for diphtheria toxin[11,12] and 613 for pseudomonas toxin;[13] hence determination of their structures presents a project of considerable technical difficulty. In this review, we give a current account of the crystallographic odyssey in search of the structures and mechanisms of these two proteins.

II. THE CRYSTALS

A pivotal contribution was made in 1980 to 1981 when John Collier succeeded in producing crystals of both toxins suitable for high-resolution structural studies. The hand of fate favored pseudomonas toxin with well-formed monoclinic crystals, space group $P2_1$, a = 60.6 Å, b = 100.2 Å, c = 59.8 Å, β = 98.6°, with one molecule per asymmetric unit.[14] Four crystal forms of diphtheria toxin were grown, the most favorable of which was a triclinic crystal, typically growing with a plate-like morphology, space group P1, a =

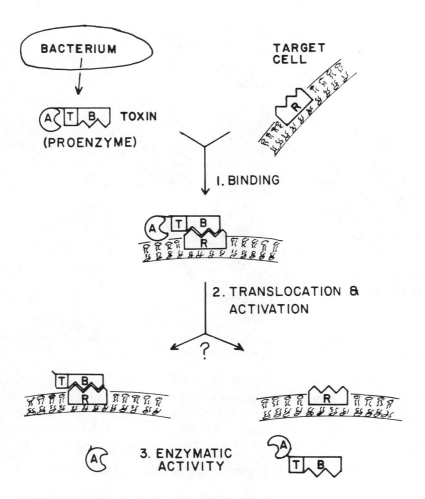

FIGURE 1. Schematic drawing of the overall pathway of toxicity of toxins such as pseudomonas and diphtheria toxins.

$$NAD^+ + EF\text{-}II \rightleftharpoons NICOTINAMIDE + EF\text{-}II$$

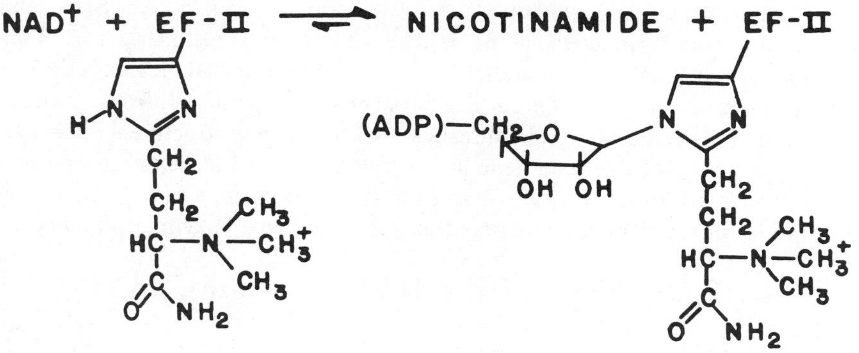

FIGURE 2. ADP-ribosylation of EF-2 as it is catalyzed by pseudomonas and diphtheria toxins.

**MEMBRANE
TRANSLOCATION**

**ENZYMATIC
ACTIVITY**

**RECEPTOR
BINDING**

FIGURE 3.   The three-dimensional structure of pseudomonas toxin, showing the alpha carbon trace in the center, and schematic drawings of the three separate structural domains, along with their assigned biochemical function.

70.8 Å, b = 70.7 Å, c = 65.3 Å, α = 95.2°, β = 91.3°, γ = 99.7°, with two molecules (possibly a specific dimeric form of the toxin) per asymmetric unit.[15] Consequently, the structure determination of diphtheria toxin has been encumbered with more inherent technical difficulties than the structure determination of pseudomonas toxin, and although significant progress has been made,[16] construction of a molecular model has not yet been completed.

III. STRUCTURE OF PSEUDOMONAS TOXIN

The structure of pseudomonas toxin was originally reported to 3.0 Å resolution,[9] and is being refined and extended with data to 2.5 Å resolution. The most striking initial observation was that the molecule has three distinct and very different structural domains (Figure 3). Although the functional boundaries of the domains have not yet been delineated precisely, the structural boundaries can be defined as: domain I, residues 1 to 253 and 364 to approximately 400, a large, antiparallel beta structure; domain II, residues 253 to 364, a helical structure protruding from the surface of the protein; domain III, from approximately residue 400 through the carboxy-terminal residue 613, a complex structure with an extended cleft. Deletion mapping of the toxin gene has revealed domain I to be the receptor-binding domain and domain III to be the ADP-ribosyl transferase domain, while domain II appears to be responsible for facilitating membrane translocation of the enzymatic activity.[10]

IV. THE QUESTION OF ENZYMATIC MECHANISM

The fact that both toxins, when activated, catalyze the identical ADP-ribosylation reaction, with indistinguishable kinetic constants, suggests they would probably share a common substrate-recognition site and enzymatic mechanism. This appears in direct conflict, however, with the reported lack of strong sequence homology of the two proteins.

The crystals of the intact toxins are of the proenzyme forms of the molecules, the form in which they are synthesized and secreted from the bacteria, prior to endocytosis into a target cell and enzymatic activation. With these crystals, we are currently constrained to studying enzymology with a protein structure that is in a proenzyme conformation. The magnitude of the conformational changes involved in activating the toxins is not known. *In vitro,* activation of pseudomonas toxin requires relatively harsh conditions: denaturation in 4 *M* urea combined with reduction of disulfide bonds.[17] Diphtheria toxin can be activated with much gentler conditions: cleavage with trypsin and the reduction of one disulfide.[18] The conditions for diphtheria toxin would suggest only modest shifts in conformation, and possibly ''deblocking'' of steric interference of active site accessibility by nonenzymatic parts of the molecule, are needed for activation. Although the *in vitro* activation protocols would not suggest this, the three-dimensional structure of pseudomonas toxin is consistent with similar requirements for its activation.

Ideally, one would like to observe substrate or substrate analog binding to the molecule crystallographically. The observation (on diphtheria toxin) that substrates are bound sequentially, with NAD$^+$ binding first and EF-2 second,[5] implies that NAD$^+$ is the ligand of choice for binding studies. However, pseudomonas toxin crystals soaked in high concentrations of NAD$^+$ and nonhydrolyzable NAD$^+$ analogs—as high as 10 m*M*—fail to yield difference Fourier maps that can be interpreted as a specifically bound ligand.[19]

Defeat in locating the intact NAD$^+$ molecule crystallographically conduced a retreat to attempting to locate NAD$^+$ fragments. It proved possible to get interpretable difference Fouriers with adenosine, and with lower signal-to-noise, suggesting lower occupancy, with AMP and ADP.[19] The adenosine binding site has been delineated with crystallographic data to 2.7 Å resolution; AMP and ADP have been observed at 6.0 Å as difference peaks overlapping that of adenosine. These results, combined with the observation that adenosine is a competitive inhibitor of the ADP-ribosyl tranferase activity,[6] support the suppostition that the adenosine binding site observed represents, to a good approximation, the binding site of the adenosine moiety of NAD$^+$. The results are also consistent with at least minor conformational changes in the toxin being required for productive binding of NAD$^+$.

These results thus fail to define the complete NAD$^+$ binding site, but they do allow one to build possible models of NAD$^+$ binding, with the adenosine moiety tethered to its observed position, which can be tested, at least indirectly, by site-specific mutagenesis.

The paradox of apparent lack of overall sequence homology between pseudomonas and diphtheria toxins, despite their identity in enzymatic activity, has been resolved by the demonstration that the two proteins have a limited sequence homology, 40% identity if lysine and arginine are equated, over a short stretch of 60 amino acids in their enzymatic domains (Figures 4 and 5).[19] The significance of the homology is apparent when the homologous residues are delineated on the three-dimensional structure (Figure 6): the majority of the residues line the enzymatic active site cleft. This weak sequence homology translates into a strong similarity in the enzymatic recognition sites of the two molecules, manifesting their shared necessity to recognize and act catalytically upon identical substrates.

The homology further lends credibility to (not to be confused with definitive proof of) the suggested model for NAD$^+$ binding; the two molecules would share several proposed specific interactions with NAD$^+$, including stacking of the nicotinamide ring on Trp 467, stacking of the adenosine ring with Tyr 481, and ionic interactions of the NAD$^+$ phosphates with Arg 458 and Arg 466.[19]

V. THE QUESTION OF MEMBRANE TRANSLOCATION MECHANISM

When one examines the sequences of the central regions of the molecules thought to be responsible for facilitation of membrane translocation, pseudomonas and diphtheria toxins are again more striking in differences than in similarities. Hydropathy profiles show diph-

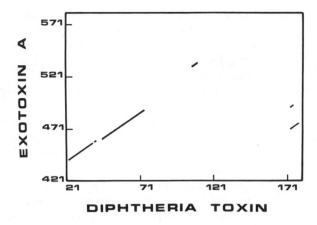

FIGURE 4.   Contour plot of sequence homology search between
the enzymatic domains of pseudomonas toxin (exotoxin A) and
diphtheria toxin. The search procedure computes the number of
identical or equivalent residues within a window of overlapping
peptides. Window size: 41 amino acids; numbers on axes are sequence numbers of center amino acid of the window. Contours are
plotted 4 sigma (4 times 1.8 amino acids) above the mean (3.3
amino acids) homology for the 41 residue window. (From
Brandhuber, B. J., Allured, V. S. Falbel, T. G., and McKay, D.
B., *Proteins*, 3, 146, 1988. With permission.)

theria toxin to be relatively hydrophobic in this region,[11,12,20] while pseudomonas toxin is not.[9,13]

The fact that the helices of domain II are not strongly hydrophobic argues against a membrane translocation mechanism that requires insertion of helices into the membrane. By implication, this argues against formation of specific transmembrane channels by the toxin.

An alternative suggestion — at this point, a suggestion without experimental corrobor-ation — would be that the helices of domain II facilitate membrane translocation by a relatively nonspecific membrane disruption mechanism. The magnitude of their hydrophobic moment places several of the helices in or near the "surface-seeking" category (Figure 7).[21] This suggests a plausible scenario in which the low-pH environment of the endosome triggers a conformation transition in the toxin molecule which "untethers" the helices of domain II and allows them to disrupt the endosomal membrane. Whether such a scenario actually takes place must await experimental scrutiny.

VI. FUTURE PROSPECTS

It appears inevitable, given the dramatic contrast between the apparent differences in primary structure between pseudomonas and diphtheria toxins, yet the fundamental similarity in their overall strategy for cell invasion, that we will learn a lot from comparison of the structures of the two molecules, once both structures are available. It further appears in-evitable that many questions will remain unresolved until we are able to compare the two structures.

Many questions regarding the ADP-ribosyl transferase mechanism will remain unan-swered until structures of the enzyme fragments in their active conformations are determined. Although no crystals are currently available for this task, the enzymatic domain of diphtheria toxin can be produced readily from the intact toxin,[8] and recently, recombinant producers of the functional enzymatic domain of pseudomonas toxin have been constructed,[10] providing avenues of approach for crystallization of the active enzymatic domains of both toxins.

```
DT    GLY ALA ASP ASP VAL VAL ASP SER SER LYS
ETA   GLU ARG LEU LEU GLN ALA HIS ARG GLN LEU
```
$_{10}$
$_{429}$

```
DT    SER PHE VAL MET GLU ASN [PHE] SER SER [TYR]
ETA   GLU GLU ARG GLY TYR VAL [PHE] VAL GLY [TYR]
```
$_{20}$
$_{439}$

```
DT    [HIS GLY THR] LYS PRO GLY TYR VAL ASP [SER]
ETA   [HIS GLY THR] PHE LEU GLU ALA ALA GLN [SER]
```
$_{30}$
$_{449}$

```
DT    [ILE] GLN LYS [GLY] ILE GLN [LYS] PRO [LYS SER]
ETA   [ILE] VAL PHE [GLY] GLY VAL [ARG] ALA [ARG SER]
```
$_{40}$
$_{459}$

```
DT    GLY THR GLN GLY ASN TYR [ASP] ASP ASP [TRP]
ETA   .....(GLN ASP LEU)..... [ASP] ALA ILE [TRP]
```
$_{50}$
$_{466}$

```
DT    [LYS GLY PHE TYR] SER THR ASP ASN LYS TYR
ETA   [ARG GLY PHE TYR] ILE ALA GLY ASP PRO ALA
```
$_{60}$
$_{476}$

```
DT    ASP [ALA] ALA [GLY TYR] SER VAL [ASP] ASN [GLU]
ETA   LEU [ALA] TYR [GLY TYR] ALA GLN [ASP] GLN [GLU]
```
$_{70}$
$_{486}$

```
DT    ASN PRO LEU SER [GLY LYS]
ETA   PRO ASP ALA ARG [GLY ARG]
```
$_{76}$
$_{492}$

FIGURE 5.   Sequence alignment of enzymatic domains of diphtheria toxin (DT) and pseudomonas toxin (ETA), with lysine and arginine considered equivalent. (From Brandhuber, B. J., Allured, V. S., Falbel, T. G., and McKay, D. B., *Proteins*, 3, 146, 1988. With permission.)

The question of membrane translocation mechanism has suffered from the lack of a reliable, reproducible *in vitro* assay of translocation activity. Until this obstacle is overcome, the mechanism(s) by which these toxin molecules facilitate translocation of an enzymatic domain across a target membrane may remain obscure.

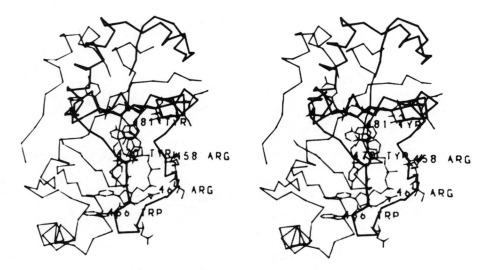

FIGURE 6.   The enzymatic domain of pseudomonas toxin: residues homologous between diphtheria toxin and pseudomonas toxin are shown superimposed on the alpha carbon backbone. The proposed model for NAD$^+$ binding is shown, in which the adenosine moiety is placed on the observed adenosine binding position. Residues which may have specific interactions with NAD$^+$ are numbered. Breaks in the polypeptide chain are as described in Reference 9.

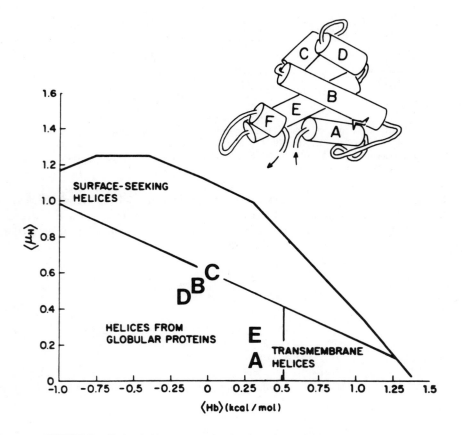

FIGURE 7.   Hydrophobic moment plot showing where the helices of domain II lie.

REFERENCES

1. **Sandvig, K. and Olsnes, S.,** Diphtheria toxin entry into cells is facilitated by low pH, *J. Cell Biol.,* 87, 828, 1980.
2. **Farahbakhsh, Z. T., Baldwin, R. L., and Wisnieski, B. J.,** Effect of low pH on the conformation of Pseudomonas exotoxin A, *J. Biol. Chem.,* 262, 2256, 1987.
3. **Collier, R. J.,** Effect of diphtheria toxin on protein synthesis: inactivation of one of the transfer factors, *J. Mol. Biol.,* 25, 83, 1967.
4. **Iglewski, B. H. and Kabat, D.,** NAD-dependent inhibition of protein synthesis by *Pseudomonas aeruginosa* toxin, *Proc. Natl. Acad. Sci. U.S.A.,* 72, 2284, 1975.
5. **Chung, D. W. and Collier, R. J.,** The mechanism of ADP-ribosylation of elongation factor 2 catalyzed by fragment A from diphtheria toxin, *Biochim. Biophys. Acta,* 483, 248, 1977.
6. **Lory, S. and Collier, R. J.,** Expression of enzymic activity by exotoxin A from *Pseudomonas aeruginosa, Infect. Immun.,* 28, 494, 1980.
7. **Middlebrook, J. L. and Dorland, R. B.,** Response of cultured mammalian cells to the exotoxins of *Pseudomonas aeruginosa* and *Cornybacterium diphtheriae:* differential cytotoxicity, *Can. J. Microbiol.,* 23, 183, 1977.
8. **DeLange, R. J., Drazin, R. E., and Collier, R. J.,** Amino-acid sequence of fragment A, an enzymically active fragment from diphtheria toxin, *Proc. Natl. Acad. Sci. U.S.A.,* 73, 69, 1976.
9. **Allured, V. S., Collier, R. J., Carroll, S. F., and McKay, D. B.,** Structure of exotoxin A of *Pseudomonas aeruginosa* at 3.0-angstrom resolution, *Proc. Natl. Acad. Sci. U.S.A.,* 83, 1320, 1986.
10. **Hwang, J., FitzGerald, D. J., Adhya, S., and Pastan, I.,** Functional domains of pseudomonas exotoxin identified by deletion analysis of the gene expressed in *E. coli, Cell,* 48, 129, 1987.
11. **Ratti, G., Rappouli, R., and Giannini, G.,** The complete nucleotide sequence of the gene coding for diphtheria toxin in the corynephage omega (tox⁺) genome, *Nucleic Acids Res.,* 11, 6589, 1983.
12. **Greenfield, L., Bjorn, M. J., Horn, G., Fong, D., Buck, G. A., Collier, R. J., and Kaplan, D. A.,** Nucleotide sequence of the structural gene for diphtheria toxin carried by the corynebacteriophage, *Proc. Natl. Acad. Sci. U.S.A.,* 80, 6853, 1983.
13. **Gray, G. L., Smith, D. H., Baldridge, J. S., Harkins, R. N., Vasil, M. L., Chen, E. Y., and Heyneker, H. L.,** Cloning, nucleotide sequence, and expression in *Escherichia coli* of the exotoxin A structural gene of *Pseudomonas aeruginosa, Proc. Natl. Acad. Sci. U.S.A.,* 81, 2645, 1984.
14. **Collier, R. J. and McKay, D. B.,** Crystallization of exotoxin A from *Pseudomonas aeruginosa, J. Mol. Biol.,* 157, 413, 1982.
15. **Collier, R. J., Westbrook, E. M., McKay, D. B., and Eisenberg, D.,** X-ray grade crystals of diphtheria toxin, *J. Biol. Chem.,* 257, 5283, 1982.
16. **Eisenberg, D.,** personal communication, 1987.
17. **Chung, D. W. and Collier, R. J.,** Enzymatically active peptide from the adenosine diphosphate-ribosylating toxin of *Pseudomonas aeruginosa, Infect. Immun.,* 16, 832, 1977.
18. **Gill, D. M. and Dinius, L. L.,** Observations on the structure of diphtheria toxin, *J. Biol. Chem.,* 246, 1485, 1971.
19. **Brandhuber, B. J., Allured, V. S., Falbel, T. G., and McKay, D. B.,** Mapping the enzymatic active site of *Pseudomonas aeruginosa* exotoxin A, *Proteins,* 3, 146, 1988.
20. **Lambotte, P., Falmagne, P., Capiau, C., Zanen, J., Ruysschaert, J. M., and Dirkx, J.,** Primary structure of diphtheria toxin fragment B: structural similarities with lipid-binding domains, *J. Cell Biol.,* 87, 837, 1980.
21. **Eisenberg, D., Schwarz, E., Komaromy, M., and Wall, R.,** Analysis of membrane and surface protein sequences with the hydrophobic moment plot, *J. Mol. Biol.,* 179, 125, 1984.

Chapter 3

TRAFFICKING OF DIPHTHERIA TOXIN IN MAMMALIAN CELLS

Catharine B. Saelinger

TABLE OF CONTENTS

I. INTRODUCTION

Diphtheria has been recognized as a clinical entity for over 100 years. Loeffler[1] first suggested that the disease was associated with the production of a diffusable material. He noted that the causative organism *Corynebacterium diphtheriae* could be isolated only from the site of infection, but that there was obvious pathology in several organs. Subsequently Roux and Yersin[2] showed that laboratory animals challenged with cell-free culture filtrates developed a fatal disease with lesions in organs similar to those seen in an infection, and von Behring and Kitasato[3] showed that antibodies directed specifically against the toxin afforded protection against infection with the organism itself. Last, development of a toxoid form of the toxin which could be used in immunization in man resulted in significant reduction of the disease throughout the world.

In the last 20 years, the protein toxin produced by *C. diphtheriae* has been extensively studied, both to determine its contribution to the disease process and to understand its molecular mechanism of action. It has been well characterized both functionally and structurally and is now considered the prototype of a growing group of toxins which are able to catalyze the ADP-ribosylation of a substrate in mammalian cells and thus exert a biological effect.[4]

II. DIPHTHERIA TOXIN BIOCHEMISTRY

Diphtheria toxin is produced only by strains of *C. diphtheriae* which are lysogenic for corynebacteriophage β or a related phage carrying the *tox* structural gene. Expression of the toxin gene is regulated by iron and high yields of toxin are produced only after the iron supply in the growth medium is depleted.[4] The toxin is secreted by the microorganism as a 535 amino acid (M_r 58,342) ADP-ribosylating toxin. Treatment with trypsin or other serine proteases followed by reduction cleaves the molecule into an N-terminal A fragment, containing 193 amino acids, which has enzymatic activity, and a C-terminal B fragment, consisting of 342 amino acids. The B region is required for binding to mammalian cells and for entry of fragment A. Fragment A catalyzes the NAD-dependent ADP-ribosylation of elongation factor 2, resulting in inhibition of protein synthesis in eukaryotic cells[5,6] (see Chapter 1). Several factors have aided in the study of the structure and function of diphtheria toxin. Among these are the isolation of mutant phages with defects in the structural gene for toxin. These include mutants which produce immunologically fully active molecules or CRMs (cross-reacting materials), which have a molecular weight of 58,000 but are nontoxic, and production of CRMs with lower molecular weights, which represent fragments of the toxin molecule. More recently, both truncated[7] and full-length mutant forms of diphtheria toxin[8] have been expressed in *Escherichia coli*. This methodology will allow expansion of studies on toxin structure and function.

III. BINDING OF DIPHTHERIA TOXIN TO MAMMALIAN CELLS

Intoxication of cells by diphtheria toxin involves several steps. Intoxication is initiated by binding of toxin to specific receptors on the cell surface; this is followed by endocytosis of bound toxin, movement through several intracellular compartments, acidification of intracellular vesicles, and delivery of fragment A to the cytoplasm where it stops protein synthesis. How diphtheria toxin crosses an intracellular membrane barrier to reach the cytosol is still not known. There is considerable evidence in the literature that diphtheria toxin binds to specific receptors on the cell surface (for review see Reference 9). Middlebrook et al.[10] demonstrated specific, saturable binding of ^{125}I-diphtheria toxin to diphtheria toxin-sensitive Vero and BS-C-1 cells growing in monolayer culture. They determined that there are approximately 150,000 toxin receptors on Vero cells, and that this represents a single class of receptor. The K_a for binding to Vero cells is 10^{-9} M. Divalent and trivalent cations are required for optimal binding.[11] Schaeffer et al.[12] found that the binding of ^{125}I toxin to cells is dependent on cell density; the estimated number of receptors on low density cells (approximately two thirds monolayer) is almost 20-fold greater than on confluent monolayers,[12] and several fold higher than reported in previous studies.[10,11] Low density cells also are more sensitive to diphtheria toxin than are high density cells; this suggests that the increased number of receptors represents functional receptors for toxin. Vero cells grow as a very compact monolayer, with cells in intimate contact with each other;[126] thus it is possible that in complete monolayers, all receptors are not accessible to the toxin. These results emphasize the importance of considering cell density when measuring binding of toxin, or any other ligand, to cells growing in monolayer culture. In addition, by electron microscopy, biotinyl-diphtheria toxin has been shown to bind specifically to Vero cells and to BS-C-1 cells.[13] Cells of lesser toxin sensitivity have too few receptors which are of too low affinity to accurately measure binding of toxin.

Information has been obtained on the nature of the interaction of diphtheria toxin with its receptor. Compounds which have several phosphate residues, such as inositol hexa-phosphate or nucleotide triphosphates, inhibit the cytotoxic effects of the toxin and prevent binding of the toxin molecule to the receptor on Vero cells.[10,14] Lory and Collier[15] have reported that ATP binds noncovalently to diphtheria toxin, and Proia et al.[16] have shown that several different nucleotides and polyphosphates are able to block the interaction of toxin with a solubilized diphtheria toxin-binding cell surface glycoprotein. They suggest that the nucleotide blocks interaction with cells by binding to the toxin molecule and not to the receptor.

There is general agreement that diphtheria toxin binds to cells via the B fragment. Studies on the interaction of altered toxin molecules with cells support this. CRM 197 is a product of a single missense mutation in the fragment A region that results in a molecule which is enzymatically inactive. Although CRM 197 is nontoxic, it is able to bind to the diphtheria toxin receptor and to block toxicity.[17] Interestingly, CRM 197 exhibits a higher affinity for cells than does wild-type toxin. Nicked CRM 197 exhibits a higher affinity for cells than does wild-type toxin. Nicked CRM 197 is far more effective than intact CRM 197 in blocking cytotoxicity of diphtheria toxin for Vero cells or in blocking binding of ^{125}I-diphtheria toxin to these cells. In contrast to native toxin, ATP does not affect binding of CRM 197. Thus the altered binding properties of CRM 197, which has only a single amino acid change in the A region, must be due to an alteration in fragment A; this in turn suggests fragment A of diphtheria toxin has a role in toxin-cell interaction.[18]

Recently, Hu and Holmes[19] reported that the insertion into lipid membranes of CRM 197 fragment A was different than that of the A domain of native toxin. These authors suggest that the single amino acid change in CRM 197 results in a molecule of different conformation than native toxin which may facilitate the entry of fragment A into membranes; they do not feel that the A fragment binds directly to receptor. Papini et al.[20] found that CRM 197 interacts more tightly with lipids than does native toxin; CRM 197 interacts with fatty acid chains of phospholipids as well as with the surface of the lipid bilayer.

A variety of treatments alter the response of Vero cells to diphtheria toxin and thus hint at factors involved in binding. Cells treated with phospholipase C or D or with proteases show reduced binding of diphtheria toxin.[21,22] Treatment with compounds which inhibit anion transport (4-acetamido-4'-isothiocyanostilbene-2,2' disulfonic acid, SITS; 4,4'-diiso-thiocyanostilbene-2,2'-disulfonic acid, DIDS) also block toxin binding.[23,24] These treatments act as effectively on metabolically inactive as active cells,[23] which suggests that they act directly on the binding site for the toxin. Other treatments, however, which alter phospho-rylation in cells also alter the ability of cells to bind toxin; these include potassium depletion of cells,[25] and incubation with salicylates, fluoride, vanadate, and phorbol esters.[23]

The physical nature of the diphtheria toxin receptor has not been fully defined. It has been described as a phospholipid, a peripheral membrane protein, and an integral membrane glycoprotein. Alving et al.[26] showed that diphtheria toxin is able to bind to the phosphate portion of several phospholipids and liposomes and that this binding is blocked by ATP and UTP. These authors, therefore, suggested that the cellular receptor for diphtheria toxin may be a membrane phospholipid. In contrast, Boquet and Duflot[27] presented information suggesting that the phosphate binding site (P) does not participate in binding diphtheria toxin to liposomes. Chin and Simon[28] partially purified a peripheral membrane protein from rabbit liver plasma membrane which blocks binding of ^{125}I-diphtheria toxin to Vero cells and prevents its cytotoxicity in these cells. In addition, ^{125}I-diphtheria toxin binds to this protein.

Eidels and colleagues have looked most extensively at diphtheria toxin binding moieties. They have isolated and characterized high molecular weight (>100,000) glycoproteins from hamster lymph node and thymus cells[29] and from guinea pig lymph node cells[30] which have specificities identical to those of the physiological receptor on the cell's surface. These glycoproteins bind to diphtheria toxin, to fragment B, and to CRM 197, but not to diphtheria toxoid or fragment A.[29-31] The binding of native toxin and fragment B to these isolated glycoproteins is inhibited by polyphosphates and nucleotides. The ability of nucleotides and polyphosphates to protect intact cells and to block toxin binding to intact cells and to purified membrane glycoproteins is similar.[10,14,31] As seen with intact cells, nucleotide-free toxin, but not toxin with a bound nucleotide, binds to these solubilized membrane glycoproteins.[31] This and other information has made these solubilized membrane glycoproteins potential candidates for the physiological diphtheria toxin receptor.[9]

Cieplak et al.[32] have identified specific diphtheria toxin binding proteins on the surface of Vero and BS-C-1 cells. Diphtheria toxin is bound to the surface of iodine-labeled cells, the cells are solubilized with a nonionic detergent, and the complex immunoprecipitated with antidiphtheria toxin; this yields a group of diphtheria toxin-binding proteins with molecular weights in the range of 10,000 to 20,000. Similar molecular weight proteins are identified when diphtheria toxin is cross-linked to cells and the diphtheria toxin-protein complex isolated. These toxin binding proteins are removed by trypsin treatment of Vero cells. Based on available data, these low molecular weight cell surface proteins may represent the functional diphtheria toxin receptors (or component of receptors) on Vero and BS-C-1 cells (i.e., they may be responsible for entry of the toxin molecules ultimately responsible for cell death). Eidels and colleagues also have looked at diphtheria toxin receptors on a hamster-derived cell line (Chinese hamster ovary K1; CHO-K1) which is of moderate diphtheria toxin sensitivity. They[33] reported that CHO-K1 cells treated with tunicamycin are less sensitive to diphtheria toxin than untreated cells, and that an acid shock treatment does not fully restore toxin sensitivity to tunicamycin-treated cells. This suggested to the authors that the functional receptor on CHO cells for diphtheria toxin is a cell surface glycoprotein. Cieplak et al.[32] noted that the chemical cross-linking methodology applied to CHO-K1 cells or hamster thymocytes gave indistinct patterns upon SDS-PAGE of the cell surface cross-linking products. This fact plus the inability of others[10,34] to show high levels of specific diphtheria toxin binding to hamster-derived cells suggest that binding to these cells is not medicated by specific receptors and that toxin entry may be by a different pathway than in Vero or BS-C-1 cells. This could imply that the search for the functional toxin receptor should be carried out only with cells of high toxin sensitivity.

Olsnes and colleagues have presented data to show that the diphtheria toxin receptor on Vero cells might be identical to or related to the anion antiport system.[35,36] The anion antiporter is thought to be the structural equivalent of the band III glycoprotein (molecular weight 100,000) which is found in erythrocyte membranes. The relationship of the smaller molecular weight proteins described by Eidels and colleagues[32] and the antiport system is unknown.

Binding of toxin to the cell may be more complex than binding to a cell surface receptor. Mekada and colleagues[37] have identified two substances in the membrane fraction of Vero cells which bind diphtheria toxin. The first is a 14.5 kDa protein, which they call receptor or receptor component; the second is an RNAse-sensitive material which blocks cytotoxicity and is called inhibitor. The inhibitor is associated with both diphtheria toxin-sensitive and -resistant cells, while the receptor is found only on sensitive cells. Receptor, but not inhibitor, binds to CRM 197. The authors suggest that the inhibitor acts as a secondary receptor in cells that are sensitive to diphtheria toxin, either by bringing toxin into closer contact with the membrane, causing a change in the conformation of the toxin molecule so that a hydrophobic region is exposed, or assisting in translocation. The concept of a secondary receptor is not unique to diphtheria toxin but also has been proposed by Montecucco[38] for botulinum and tetanus toxins. The receptor described by Mekada is similar in molecular weight to that described by Cieplak et al.[32] and due to its binding characteristics is considered to be a component of the functional diphtheria toxin receptor.

Lastly, Creagan et al.[39] reported that the introduction of human chromosome 5 into resistant mouse cells rendered the cells sensitive to diphtheria toxin. More recently, the gene for toxin sensitivity was localized to the q23 region of human chromosome 5.[40] While it has been proposed that this could represent the gene for the toxin receptor, this has not been proven and requires further examination.

IV. DIPHTHERIA TOXIN INTERNALIZATION

There is agreement as to the general steps involved in diphtheria intoxication of sensitive cells: (1) binding of toxin to specific receptors, (2) entry by receptor-mediated endocytosis into an acidic prelysosomal vesicle, and (3) entry of toxin into the cytosol where elongation factor 2 is inhibited. The site(s) at which toxin is converted into an enzyme active form and from which toxin enters the cytosol has not been unequivocally determined. Nor has the mechanism by which the toxin crosses the membrane barrier been proven.

The movement of toxin in a mammalian cell can be mapped in several different manners. Since the endpoint of diphtheria toxin action is inhibition of protein synthesis, protocols with known sites of action can be tested to determine their effect on toxicity. The effect of diphtheria toxin on mammalian cells which have mutations at different points in the endocytic pathway can be assessed. Actual "movement" of toxin in a cell can be followed using electron microscopy or video-intensification microscopy. All of these approaches have been taken alone or in combination to dissect the stations diphtheria toxin passes on its journey in the cell.

There are several problems that need to be considered in discussing toxin movement within cells. First, the criticism can always be raised as to whether or not one is following the toxin molecules actually involved in killing the cell. Although greater than 100,000 diphtheria toxin molecules bind to cells (under saturating conditions), not all of these molecules enter the cell once internalization is initiated.[13] It has been shown that only a small percentage of the total number of diphtheria toxin molecules which bind to the surface of a sensitive cell actually participate in the inactivation of elongation factor 2.[41] Furthermore, Yamaizumi et al.[42] have calculated that one single molecule of diphtheria toxin introduced directly into the cytoplasm theoretically is capable of blocking cell division. For these reasons, it is technically difficult to differentiate the toxin molecules following the putative pathway (i.e., leading to inhibition of protein synthesis) as opposed to those (the majority) which are presumably routed to lysosomes where they are degraded, and thus do not participate in inhibition of protein synthesis. Second, biochemical methodologies measure an endpoint; they measure the contribution of toxin molecules which enter cells by nonreceptor-mediated routes and eventually aberrantly reach the cytoplasm to stop protein synthesis, as well as measuring the activity of toxin molecules entering cells by the productive route. Third, ultrastructural studies, although highly reproducible, examine only a very small population of cells. In addition, the methodologies used to identify ligand may result in the ligand being routed in a manner different from native protein.[43,44]

What is known about the steps in the entry and subsequent intracellular movement of diphtheria toxin? The first step after toxin binds to a sensitive cell, such as Vero or BS-C-1, is movement of toxin, presumably coupled to receptor, to coated pits. Ultrastructural studies have shown that biotinyl-diphtheria toxin (identified with colloidal gold) binds randomly to the surface of BS-C-1 or Vero cells at 4°C.[13] Immediately upon warming to 37°C, the toxin moves to coated areas of the membrane and from there is seen intracellularly in noncoated vesicles. This rapid movement of toxin into the cell corresponds well to the rapidity with which diphtheria toxin becomes inaccessible to antitoxin neutralization.[45] Immediately after warming Vero cells to 37°C, a small percentage of prebound diphtheria toxin reaches a site at which it is no longer accessible to neutralization by diphtheria antitoxin.

Larkin et al.[46,47] found that depletion of intracellular potassium in human fibroblasts reversibly stops formation of coated pits and the receptor-mediated endocytosis of ligands, such as low density lipoprotein and epidermal growth factor. Similarly hypotonic shock and depletion of intracellular potassium in Hep-2 cells results in disappearance of clathrin-coated pits and inhibition of the receptor-mediated endocytosis of transferrin.[48] More important to this discussion, potassium depletion protects Hep-2 cells from the action of diphtheria toxin. This protection is presumed to be due to defects in endocytosis of toxin, and not to a loss of toxin receptors or a decrease in level of toxin binding to cells. The data would support the hypothesis that diphtheria toxin must enter cells via coated pits to express biological activity.

It should be pointed out that inhibition of the endocytosis of ligands internalized via coated pits is not the only effect that potassium depletion exerts in cells; many cellular processes are altered by potassium depletion.[25,49] Potassium depletion induces changes in intracellular pH. The membrane potential is reduced since the Na/K-ATPase cannot maintain the proper ion gradient. Depletion of potassium also results in changes in the cellular level of cations and of chloride.

V. ROLE OF ACIDIFIED COMPARTMENTS IN TOXIN TRAFFICKING

In the routing of physiologically relevant ligands, the next step after entry into coated pits and coated vesicles is transport to prelysosomal acidic organelles, here to be designated endosomes. There are several possible events which can occur in an acidic environment including: (1) conversion of diphtheria toxin to an enzyme active form (involving nicking and disulfide bond reduction); (2) conformational change of the toxin molecule so it can penetrate lipid bilayer membranes; and (3) entry into the cytoplasm.

To date, no definitive evidence has been obtained to show which event(s) occur in endosomes. The possible role of an acidic environment in toxicity has been approached indirectly by numerous investigators. This includes looking at the effect of acidic pH on toxin conformation, the role of pH in diphtheria toxin penetration of artificial membrane systems, acid pH-induced entry of diphtheria toxin into the cytosol directly through the plasma membrane, protection of cells by agents which increase intraorganelle pH, and use of mutant cell lines defective in endosomal acidification. Each point will be addressed separately.

A. PROTECTION OF CELLS FROM DIPHTHERIA TOXIN-INDUCED TOXICITY

Several chemicals and drugs which are known to interfere with internalization and/or degradation of physiologically relevant ligands by mammalian cells also block diphtheria toxin action. These agents have been used to dissect the intracellular movement of toxin. Kim and Groman[50,51] were the first to report that ammonium salts, including ammonium chloride, prevent the expression of diphtheria toxin cytotoxicity, possibly by inhibiting specific but not nonspecific uptake of toxin. Ivins et al.[52] and Middlebrook and Dorland[53] tested the ability of a number of chemicals to protect Hep-2 or HeLa cells from diphtheria toxin. Several energy inhibitors including sodium fluoride and salicylic acid were protective, while other energy inhibitors were without effect. Membrane perturbants and local anesthetics (lidocaine and procaine) also provided partial protection from diphtheria toxin. These studies were done using cells of moderate toxin sensitivity, where it is difficult to differentiate toxin internalized by an efficient receptor-mediated route from toxin internalized by a nonspecific mechanism.

More recent studies using monkey kidney cells (e.g., Vero) shed more light on the putative toxic pathway. In these cells it is presumed that a receptor-mediated entry process is being tracked. Dorland[54] showed that a large number of alkylamines both protect Vero cells from diphtheria toxin and block lysosomal degradation of the toxin. It is now known that these substances raise the pH of several intracellular organelles including endosomes and lysosomes. In contrast, ammonium chloride protects Vero cells from diphtheria toxin but does not block the bulk uptake and subsequent degradation of toxin; however, it does maintain a small fraction of potentially toxic molecules in a position accessible to antitoxin neutralization at 37°C.[54-56] Similarly, methylamine[13] appears to maintain toxin at a site where it is accessible to antitoxin neutralization. The protective effects of methylamine are reversible. Following removal of amine from the medium, diphtheria toxin appears to return to the cell surface, where it presumably is reinternalized. Evidence for this recycling pathway comes from the fact that addition of antitoxin to cells at 37°C, after amine removal, results in high levels of protection, while 4°C incubation with antitoxin is less effective.[13] Mekada et al.[57] also have reported a rapid loss of cell-associated ^{125}I-labeled diphtheria toxin following removal of methylamine. Thus, all available evidence suggests that diphtheria toxin is sequestered within the cell in the presence of amine, and that upon removal of the amine block, toxin returns to the surface and re-enters the cell via receptor-mediated endocytosis.

Further support for the role of an acidic environment in the entry of diphtheria toxin into the cell cytoplasm first came from the laboratories of Olsnes[58-60] and of Draper.[56] As just described, acidotropic agents which increase the pH within an acidic intracellular organelle protect cells from diphtheria toxin. Surface-bound diphtheria toxin is able to evade this block and enter cells upon incubation in low pH medium. Nicked toxin inhibits protein synthesis much more rapidly than intact toxin under the same conditions. The metabolic inhibitors, 2-deoxyglucose and sodium azide, block diphtheria toxin internalization[59,61] and this block cannot be overcome by low pH incubation.[59] In addition, low pH incubation can restore toxicity when endocytosis is blocked by incubating cells at low temperatures.[56,59] These experiments have been repeated by other investigators using a variety of acidotropic agents.[13,34,55]

B. POSSIBLE ROLE OF ACIDIFICATION

What role does the acidic pH have? Diphtheria toxin has been shown to undergo a distinct change in conformation at low pH.[62-65] At pHs below 5.3 (37°C), diphtheria toxin undergoes a rapid ($t_{1/2} < 3.5$ min.) change in conformation resulting in increased hydrophobicity and partial exposure of buried tryptophanes. This transition occurs over a narrow pH range (0.2 pH units). Fluorescence properties and circular dichroism show that this represents only an opening of the toxin molecule, not an extensive unfolding.[63] These pH-dependent conformational changes occur both in the presence and absence of membranes and appear to involve burial of certain sites in the molecule and exposure of other sites.[65] That this conformational change occurs at a pH similar to that found in endosomes and that it occurs rapidly are consistent with events hypothesized to occur in intact cells which lead to toxicity. These results also are consistent with the proposal that low pH may trigger membrane penetration by diphtheria toxin.

What evidence is there for membrane penetration at an acidic pH? Donovan et al.[66,67] and Kagan et al.[68] used artificial lipid bilayers as models to examine how the A fragment of diphtheria toxin enters the cytoplasm. In their studies, they looked at the electrochemical conductance across bilayers that occurs when diphtheria toxin or toxic fragments with hydrophobic domains are added to the bilayers. Ion-conducting channels are formed at a low pH and at a transmembrane potential that is positive on the side to which the toxin is added. Channel formation is most effective when the pH is asymmetric, i.e., toxin is added in an acidic compartment and the opposite compartment is neutral. Diphtheria toxin channel formation also depends upon the lipid composition of the membrane and specifically requires phospholipids which have negatively charged head groups.[67] Other investigators have looked at the formation of ion channels in lipid vesicles and have found that the factors which influence channel formation in planar membranes — such as pH, potential, and presence of phosphoinositides — also influence channel formation in lipid vesicles.[69]

An important question must be addressed before it can be proposed that these channels represent tunnels through which active toxin fragments enter the cytosol and that is, are the channels large enough to allow movement of the peptide chain? The initial work by Donovan et al.[66] suggested that under acidic conditions native diphtheria toxin forms anion-selective channels which are too small (5 Å) to allow passage of fragment A. In contrast, Kagan et al.,[68] working with multilamellar vesicle, found that the channels formed by the CRM 45 B fragment allow passage of nonelectrolytes, such as polyethylene glycol 1500; this suggests that the pore size is at least 18 Å in diameter and is large enough to accommodate fragment A of diphtheria toxin in an extended form.

More recently, Hoch et al.[70] using planar lipid membranes confirmed that the diphtheria toxin channel is large enough to accommodate fragment A. Charge is an important determinant in the selectivity properties of the diphtheria toxin channel. At pH 5.5, the channel prefers cations, while at pH 3.5 the diphtheria toxin channel prefers anions. These authors suggest that a portion of the B fragment interacts with lipid bilayer membranes to form channels, and that channel formation is most effective in the presence of a pH gradient compatible with the gradient found across intracellular acidic vesicle membranes. The pH gradient across the acidic vesicle may be a driving force for the translocation of fragment A; it could act by causing the A chain to unfold at the mildly acidic pH found in the vesicle, and refold at the neutral cytosolic pH. Sandvig and Olsnes[71] looked at the channels formed in Vero cell membranes associated with the translocation of diphtheria toxin. Nicked toxin was bound to cells in the cold and the cells were exposed to low pH conditions to induce toxin entry. Under these conditions, the cells were permeable to the monovalent cations, K^+ and Na^+, and to choline$^+$ and glucosamine$^+$; there was no increased permeability to SO_4^2, Cl^-, sucrose, or glucose. Formation of the cation-selective channels requires more than binding of toxin or receptor and insertion of the B fragment into the membrane; channel formation is dependent upon a transmembrane pH gradient. Thus transport of diphtheria toxin across the Vero cell membrane induces formation of cation-selective channels that have properties which are similar to the channels formed in artificial lipid membranes.[66,67]

Donovan et al.[72] took a different approach to studying translocation of toxin across membranes. Lipid vesicles are loaded with elongation factor 2 and radio-labeled NAD, then exposed to high concentrations of nicked diphtheria toxin at acidic pHs. Some of the elongation factor 2 within the vesicles becomes ADP-ribosylated, suggesting that at least the A fragment of the toxin enters the vesicle.

All studies with artificial systems have been done in the absence of proteins; this implies that the movement of toxin (or a fragment of toxin) through the membrane does not require the presence of receptor. One possible role for the receptor is to allow the toxin molecule to attach to the cell surface and direct it to coated pits and into endosomes where it encounters the proper environment for continued processing. As already discussed, acidic pH triggers a conformational change in the toxin molecule which renders it hydrophobic.[63,65]

Both membrane potential and the proton gradient across the membrane have been shown to stimulate formation of channels in planar lipid bilayers using diphtheria toxin or CRM 45.[65,68] Work in this artificial system is hard to relate to what occurs in intact cells. For this reason, Olsnes and colleagues[73] looked at the effect of changing the pH gradient and the membrane potential across the membrane of intact Vero cells on the ability of toxin to enter the cytoplasm and inhibit protein synthesis. Their results show that reduction in the membrane potential does not greatly alter toxicity. However, acidification of the cytosol inhibits toxin entry. They found that an inward directed proton gradient of at least 1 pH unit is needed for toxin entry, and suggest that this gradient may act as the driving force for the entry of fragment A into the cytosol.

Permeant ions (e.g., $Cl-$, $Br-$, $I-$, NO_3-) and free anion passage also are required for diphtheria toxin to enter the cytosol of Vero cells.[23-25] However, even when permeant anions are included in the incubation medium, diphtheria toxin does not enter cells if certain inhibitors of anion transport, such as DIDS and SITS are added. Recently, evidence has been presented for both a sodium-linked and a sodium-independent bicarbonate/chloride exchange in Vero cells.[74] These two antiporter systems are blocked by different inhibitors. Experiments by Olsnes et al. indicate that agents which inhibit the sodium-independent antiport system are most efficient in inhibiting diphtheria toxin entry; therefore, there may be a connection between the sodium-independent anion antiport and the movement of diphtheria toxin into the Vero cell cytoplasm.

VI. STUDIES USING MUTANT CELL TYPES

Another way to dissect the intracellular routing of ligands in mammalian cells is to obtain mutant cell lines which have defects in different stations along the route, and determine how these cells handle toxin. Diphtheria toxin-resistant cell lines, predominantly Chinese hamster cell lines, have been isolated and characterized in several laboratories and have been classified by the Moherings[75] into two major groups: class I mutants (DT^{R}I) and class II (DT^{R}II). Class I mutants are toxin-entry mutants. Class II mutants are translational mutants at the level of elongation factor 2 and are cross-resistant to diphtheria toxin and to pseudomonas exotoxin A. For the purposes of this review, we will be concerned with the entry mutants and how they advance our understanding of diphtheria toxin internalization and processing. Cells can be altered at several different stages in entry/processing with a resulting increase in resistance to toxin. This includes alteration in receptor, in the entry process itself, in endosomal function presumably acidification, in Golgi function, and in lysosomal function.

A. ALTERED BINDING ACTIVITY

Most diphtheria toxin mutants isolated have normal or only partial reduced binding activity. Kohno et al.,[76] however, have isolated entry level mutants of CHO cells which lack a functional receptor for diphtheria toxin (DT^R I). However, because of the low level of toxin binding, it is not possible to state definitely that these mutants totally lack receptor. The defect in these cells cannot be overcome by exposure to low pH medium. Based on these studies, the authors conclude that these are entry-deficient mutants in which the primary toxin receptors have reduced or negligible diphtheria toxin binding activity. Further, they suggest that these cells lack "high affinity" receptors, but have "low affinity" (secondary) receptors for diphtheria toxin. Kaneda et al.[77] have prepared substances from both toxin-sensitive and toxin-resistant cells which competitively inhibit diphtheria toxin binding, and which may be the "secondary" receptors.

TABLE 1
Characterization of Mutant Cell Lines Which Are Resistant to
Diphtheria Toxin

Line	Diphtheria toxin	Pseudomonas exotoxin A	Viruses	Ref.
KB (wt)	S	S	S	78,79
KB-R2A	R	ND[a]	R	78,79
CHO (wt)	S	S	S	
G7.1	R[b]	R[b]	ND	80
D PV[r]	R	R	R	34,83,84
Dip[r]	R	S	S	34
PV[r]	S	R	R	83
DTG 1-5-4	R	R	R	89
DTF 1-5-1	R	S	ND	85,89
GEI	R	R	S	90

S, sensitivity to toxin or to virus is comparable to wild type cell; R, increased resistance to toxin or to virus, as compared to wild type cells.

[a]   Not determined.
[b]   At nonpermissive temperature.

B. ALTERED ENTRY/ACIDIFICATION

There are a number of mutant cell lines lacking components of the acidification process which exhibit blocks in the productive or toxic transport of diphtheria toxin (Table 1). The Moehrings have isolated a KB cell variant (KB-R2A) that is resistant to diphtheria toxin and to infection by RNA viruses[78] and exhibits altered degradation of epidermal growth factor.[79] The initial steps in hormone entry are normal, i.e., clustering into coated pits and movement to endocytic vesicles. In addition, the early processing of epidermal growth factor in endosomes is similar in mutant and wild-type cells; however, a later intracellular processing step is inhibited. The defect could be due to reduced fusion of endosomes with lysosomes, inhibition of lysosomal acid hydrolases, or decreased levels of intracellular acid hydrolases.[79]

Marnell et al.[80] derived a CHO cell mutant (G.7.1.) that has a temperature-sensitive conditional lethal lesion affecting vesicle acidification; this defect is associated with non-lysosomal vesicles including endosomes, but is not seen in secondary lysosomes.[81] The mutation which is responsible for the defective acidification seen in G.7.1 cells is believed to lie at the level of the proton pumps present in coated vesicle-endosomal compartments.[82] At the nonpermissive temperature (39.5°C), G.7.1 cells are resistant to diphtheria toxin, pseudomonas exotoxin A, and modeccin, as well as exhibiting defects in transferrin-mediated iron uptake. Thus, defects in acidification in early steps of the endocytic pathway may result in improper routing/processing of toxin molecules.

Extensive characterization of CHO cell mutants has been carried out by the Moehrings.[34,83,84] They have derived one class of mutant cells which is cross-resistant to diphtheria toxin, pseudomonas exotoxin A, and several viruses including sindbis virus, Semliki forest virus (SFV), and vesicular stomatitis virus[34] (DPV[r]). The resistance to diphtheria toxin can be overcome by exposing cells to low pH.[34] These cells have been shown to have a defect in the ATP-dependent acidification of endosomes, but not of lysosomes. Although there is a defect in endosome acidification, it is not total and isolated endosomes from mutant cells have an internal pH nearly one pH unit below that of the surrounding buffer. If this level of acidification occurs in intact endosomes, then it is not enough for expression of diphtheria toxin toxicity.[84] A second class of CHO cell mutants (Dip[r]) are resistant only to diphtheria toxin and do not exhibit cross-resistance to pseudomonas toxin or to viruses. These cells have the same number of high affinity binding sites as do parent CHO cells. Toxin resistance in these cells is not overcome by exposing them to low pH; thus, they are not thought to have a defect in their ability to acidify intracellular vesicles. In addition, Dip[r] cells have a normal sensitivity to CRM 45. It is hypothesized that Dip[r] cells are resistant to diphtheria toxin due to a block in a step occurring between binding of toxin to receptors and the pH dependent step required for toxicity.[34] The authors hypothesize that these cells may be unable to cleave a peptide bond in fragment B, or that the receptor itself is not in the proper conformation required for insertion of the toxin molecule into the target membrane. In addition, no one has examined these mutants to determine if they internalize diphtheria toxin via coated pits or not. The overall result is that these cells cannot effectively transport the A fragment of diphtheria toxin into the cytosol. A third group of mutants, PV[r], are sensitive to diphtheria toxin, but resistant to pseudomonas exotoxin and certain enveloped viruses.[83]

Robbins et al.[85] also have isolated CHO cell mutants which have some similarity to the DPV[r] mutants. These mutants exhibit pleiotropic defects in receptor-mediated endocytosis; the cells show decreased accumulation of iron from tranferrin, decreased uptake of lysosomal hydrolases, increased resistance to certain enveloped RNA viruses, and increased resistance to diphtheria toxin. All of the mutants are not the same and differential uptake of certain ligands is noted (see Table 1). For example, DTG 1-5-4 exhibits increased resistance to several toxins (pseudomonas exotoxin A, diphtheria toxin, and modeccin) and viruses (vesicular stomatitis virus and sindbis virus) and a decreased uptake via the mannose-6-phosphate receptor.[86] In contrast, the mutations in DTF 1-5-1 do not decrease the endocytosis of modeccin, ricin, or pseudomonas exotoxin, and have only minimal effect on the uptake and degradation of low density lipoprotein.[85] It recently has been reported[87,88] that acidification of early endocytic compartments is defective in both of these mutants (DTG 1-5-4 and DTF 1-5-1); after 5 min, the average endosomal pH is 6.7 in the mutant cell line, compared to 6.3 in parent wild-type cells.[87] Acidification of large endosomes and recycling endosomes appears to be near normal in the mutant cell lines. In addition, DTG 1-5-4 has defects in certain Golgi-associated functions; for example, the processing of certain precursor viral proteins, the galactosylation and transport to the cell surface of certain sindbis virus glycoproteins are inhibited.[89] These authors speculate that the defect in acidification of early or of sorting endosomes may partially explain the resistance of these cells to diphtheria toxin. They point out, however, that the bulk of the toxin is delivered to lysosomes to be degraded, and, therefore, must pass through later prelysosomal compartments which are acidic. This acidity is not sufficient to allow expression of toxicity, supporting the concept that other cell-associated characteristics besides reduction in pH are needed to allow toxin to be active.

Recently Kohno et al.[90] have isolated a new type of mutant (GE1) which has a receptor for diphtheria toxin and normal acidification function, but is defective in some step in toxin entry. Elongation factor 2 in these mutant cells is 50% ADP-ribosylatable; however, protein synthesis is normal, even in the presence of high levels of toxin. Exposure to low pH does not render GE1 cells sensitive to toxin. Interestingly these cells are cross-resistant to pseudomonas exotoxin A, but are as sensitive as parent cells to infection with vesicular stomatitis virus. The authors suggest that an additional cellular factor besides receptor and endosome acidification is needed for intoxication and that this factor appears to be required for both pseudomonas and diphtheria toxin.

VII. SEPARATION OF BINDING AND TRANSLOCATION

Genetic manipulations have been used to separate the binding function of fragment B from the translocation function.[91-93] One mutant of diphtheria toxin, CRM 107, has point mutations at positions 390 and 525 in the carboxy-terminal portion of the B polypeptide chain which reduce toxin binding 8000-fold and reduce toxicity for Vero cells 10,000-fold.[91] This mutant, however, has normal translocation function. Linking CRM 107 to a monoclonal antibody which is specific for human T cells results in an immunotoxin which is fully toxic to antigen-positive target cells. This immunotoxin has full enzyme activity, full translocation activity, lacks native toxin binding, and has a new binding domain provided by the antibody. Generation of CRM 107 shows that the binding activity and translocation activity of diphtheria toxin can be separated.

Colombatti et al.[93] had previously tried to separate the binding and translocation functions of fragment B by domain deletion, but were not successful. They deleted a 17-kDa region from the carboxy terminus of fragment B, and used this altered fragment, intact toxin, and fragment A to make immunotoxins. The conjugate made with the intact toxin was 100-fold more toxic than the immunotoxin made with the altered toxin; the latter molecule was 100-fold more toxic than the immunotoxin prepared with fragment A. It was not possible, however, to differentiate the carboxy-terminal translocation activity from binding activity.

Taking a different approach, Hayakawa et al.[94] isolated a series of monoclonal antibodies against diphtheria toxin and used these to relate structural domains of the toxin molecule and function. One antibody which reacts with the region between 30 and 45 kDa from the amino terminus blocks entry of toxin into the Vero cell cytosol, but does not block binding. This is in contrast to a second antibody which reacts with the 17-kDa carboxy terminus, which does block toxin binding. Their data also support the theory that fragment B has two functional domains. One domain is responsible for toxin binding to the cell surface and the other domain is involved in the translocation of fragment A across the cell membrane.

Initial experiments to study toxin translocation assumed the toxin entered the cytosol from an acidic organelle, and thus mimicked this condition by exposing cells to a low pH. As already described, this procedure induced entry of surface-bound diphtheria toxin, directly across the plasma membrane. Using this method, one could measure the final effect of toxin, namely decrease of protein synthesis. It is far more difficult to dissect the movement of toxin across the membrane of an intracellular organelle; however, progress is being made in this direction.

Moskaug et al.[95] took a more direct approach to address this question. Their studies still follow toxin, bound to the surface of Vero cells; however, they localize fragment A and B in cells following exposure to low pH (pH 4.5). In this case, iodinated nicked diphtheria toxin is bound to cells, cells are exposed to pH 4.5 medium, and treated with pronase to remove cell surface material. SDS-PAGE reveals two fragments, 25 and 20 kDa, which are protected from pronase digestion after low-pH treatment, indicating they are no longer exposed on the cell surface. These fragments represent about 5 to 10% of the total radio-activity initially bound to cells. Further analysis indicated that the 20-kDa peptide has ADP-ribosylation activity (i.e., is fragment A) and the 25-kDa peptide is believed to be a cleavage product of the B fragment. Further, the 20-kDa fragment was shown to be released into the cell cytosol (was found in the soluble fraction), while the larger polypeptide remained associated with membranes (in the cell pellet). This localization is what would be expected if the smaller fragment has enzyme activity, and the larger fragment is involved in toxin binding or translocation.

Energy imput is required for membrane translocation in all other protein systems which have been studied. Energy can be provided by ATP, by the membrane potential, by the chemical proton gradient across the membrane, or by a combination of these factors. Moskaug and colleagues[95] report that the energy required for translocation of fragment A of diphtheria toxin into the cell cytosol appears to be provided by transmembrane proton gradient. When a reduction in the inward-directed proton gradient was prevented, fragment A was no longer released into the cytosol. Anion transport also may be required for translocation of diphtheria toxin as presence of anion transport inhibitors block release of fragment A. This inhibition is in agreement with the author's hypothesis that an anion transporter is part of the diphtheria toxin receptor. In contrast, they feel that neither electrical depolarization, nor ATP depletion interferes with release of fragment A. Thus ATP, in their hands, does not appear to be required for diphtheria toxin entry.

Hudson et al.[96] have reached different conclusions and report that ATP is required for translocation. These authors propose a novel model for entry of diphtheria toxin into the cell cytosol. Initially, they reported that the translocation of diphtheria toxin into the cell cytoplasm, not ADP-ribosylation of EF2, is the rate-limiting step in the inhibition of protein synthesis.[97] They proposed that toxin is released as a bolus into the cytosol and that the size of this bolus remains constant, with about 20 fragment A molecules being released from a single vesicle at a time.[98] More recently, they presented a two-phase model for intoxication.[96] In the first part, toxin is internalized in endocytic vesicles and is processed there. This vesicle maturation process requires either establishment of a plasma membrane potential or a pH gradient across a cellular membrane or both. In the second phase, the mature vesicle is now able to release A chain into the cytosol. How release or translocation occurs is not known, but it requires the presence of at least one component of the proton motive force across the membrane.

VIII. DISULFIDE BOND REDUCTION

Diphtheria toxin must be nicked and the disulfide bond holding the A and B fragments together (between cysteine residues at position 186 and 201) must be reduced to allow full enzymatic activity of the A fragment. Olsnes and colleagues[99] have examined the cell-mediated reduction of the disulfide bond, and compared these conditions with requirements for toxicity. ^{125}I-labeled nicked diphtheria toxin is bound to Vero cells and the cells are then exposed to pHs of 5.0 and below. Monensin is present to prevent toxin entry and processing by the normal route. Under these conditions, a small fraction (5 to 10%) of the cell-associated toxin is reduced to yield A and B fragments. The authors calculate that this represents a reduction of 200 to 400 toxin molecules per cell, and is in the range of the number of molecules calculated to be internalized into the cytosol.[41]

The transmembrane pH gradient has been proposed to be a driving force in the transfer of fragment A across the membrane.[73] When this proton gradient is dissipated by cytosol acidification, toxin entry is inhibited. Similar treatment also inhibits disulfide bond reduction. Several agents could be involved in the cell-mediated reduction of the toxin; these include the sulfhydryls, glutathione and cysteine, or the thioredoxin system. Since cysteine and glutathione work well at a neutral pH, as would be found in the cytosol, it could be hypothesized that under normal conditions, low pH is required only to mediate conformation changes and facilitate toxin passage through the membrane. Consistent with a cytoplasmic site for disulfide bond reduction is the work by Wright et al.[100] which showed that modifications of the cysteine disulfide which link the A and B fragments result in inhibition of a step in the intoxication process after binding. The authors speculate that the interchain disulfide bond functions in transmembrane transport of the A-chain. In addition, Blewett et al.[63] stated that disulfide bond reduction is inhibited by a low pH environment and presumably would not occur in endosomes. This statement must be reconciled with the fact that Moskaug et al.[99] carried out their *in vitro* experiments at pH 5.0 and below. If reduction of the disulfide bond between A and B fragment of diphtheria toxin, in an intact cell, actually occurs at the neutral cytoplasmic side of the membrane, then we will need to consider how the toxin is inserted into the membrane and how much of the nicked molecule passes through the membrane so that bond reduction can occur.

IX. INTRACELLULAR LOCALIZATION OF DIPHTHERIA TOXIN

The experiments using aciditropic agents or mutant cells deficient in acidification suggests that diphtheria toxin enters the cytoplasm after encountering an acidic intracellular vesicle. A more direct approach to direct this question is to localize the toxin within subcellular organelles. To this end, Fedde and Sly[101] compared the intracellular trafficking of diphtheria toxin and epidermal growth factor (EGF) in BSC-1 cells, a cell line highly sensitive to toxin. Other studies have established that EGF is internalized and subsequently degraded via the endosome-lysosome pathway (see Chapter 1). Cells were incubated with iodinated diphtheria toxin for 25 or 90 min at 37°C, surface-associated toxin was removed with trypsin-EDTA, and cells were homogenized and fractionated on a Percoll gradient. Toxin was found exclusively in a fraction of intermediate density (1.035 g) which is characteristic of endosomal vesicles; no toxin was detected in fractions containing secondary lysosomes. In contrast, iodinated EGF was localized in both the endosomal- and lysosomal-enriched fractions.

BSC-1 cells were pretreated with a thiol protease inhibitor, leupeptin, to make certain that the inability to detect toxin in lysosomes was not due to rapid toxin degradation by these organelles. Even in the presence of leupeptin, only minimal (less than 2% of the cell-associated diphtheria toxin) was seen in secondary lysosomes. Interestingly, toxin is readily degraded by cells even though it does not, by their experiments, get delivered to lysosomes. The data would suggest that degradation of diphtheria toxin does not occur in lysosomes. These authors hypothesis that toxin degradation could take place in endosomes[102] or in the cytosol by the ubiquitin pathway.[103] Endosomal degradation of albumin has been described previously.[102] These results are in contrast to work by other investigators, and further experiments will be required to determine the exact site of diphtheria toxin degradation.

X. MORPHOLOGICAL STUDIES ON TOXIN ENTRY

Morphological evidence is available on the movement of diphtheria toxin in cells. Keen et al.,[104] using video-intensification microscopy, reported that diphtheria toxin enters human W138 and mouse 3T3 cells, and by 30 min is seen in the same vesicles as α-2 macroglobulin, thus suggesting a role for receptor-mediated endocytosis in entry. Morris and colleagues[13] have provided direct evidence, on the ultrastructural level, that diphtheria toxin enters Vero and BS-C-1 cells by the receptor-mediated endocytic pathway. In these experiments, biotinyl toxin is incubated with the cell monolayer at 4°C; succinyl-avidin-gold colloids are added (4°C) to mark the toxin on the cell surface, and cells are warmed to 37°C to allow internalization of the toxin-gold complex. Biotynyl-diphtheria toxin binds to receptors randomly distributed on the Vero and BS-C-1 cell surface. In cells maintained at 4°C, less than 5% of the toxin is seen associated with coated areas. Surface binding is specific and can be reduced to near background levels by incubation in the presence of native toxin. Following warming to 37°C, toxin moves rapidly (60 s) to coated areas of the membrane and is then internalized into coated vesicles and delivered to noncoated organelles identified morphologically as endosomes. On occasion, diphtheria toxin is seen in the Golgi region of the cell between 5 and 10 min of internalization. The Golgi is difficult to identify morphologically in these cell types; thus it has proven hard to quantitate the level of this compartmentalization. At no time is diphtheria toxin seen in Golgi cisternae. With continued incubation at 37°C, toxin is found in lysosomes (identified by overnight incubation at 37°C with horseradish peroxidase (HRP), followed by 2-h inubation in HRP-free medium). Representative electron micrographs which depict intracellular movement of diphtheria toxin are presented in Figure 1. It should be stressed that the time required for movement from the cell surface to endosomes to lysosomes as determined by electron microscopic visualization[13] correlates well with the kinetics of diphtheria toxin internalization reported by Marnell et al.[45] using biochemical methodologies. Thus both available biochemical and electron microscopic data show that the putative toxin molecules are internalized by sensitive cells via clathrin-coated pits (receptor-mediated endocytosis) and enter a trafficking pathway involving prelysosomal acidic

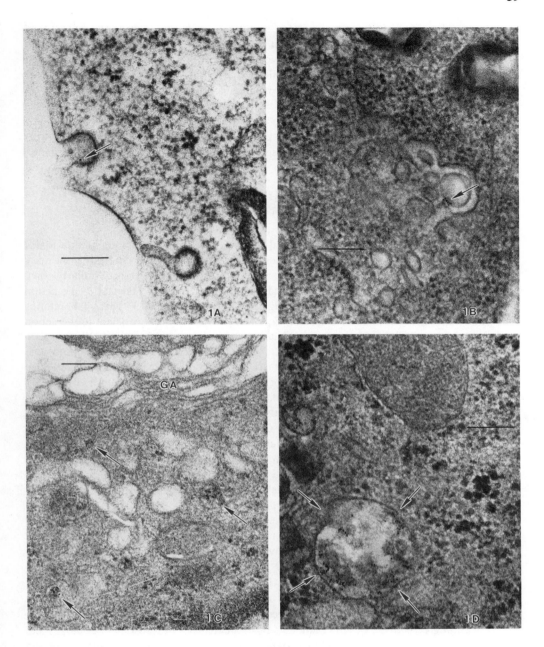

FIGURE 1. Intracellular movement of biotinyl-diphtheria toxin in Vero cells. Cell monolayers were incubated in Hank's balanced salts for 2 h, cooled to 4°C, then incubated sequentially with 150 ng/ml biotinyl-toxin and streptavidin-gold (5 nm), warmed to 37°C as indicated, then fixed and processed for electron microscopy.[12] A. 30 s at 37°C; note gold in coated pit. B. 5 min; note gold in amorphous vesicle, considered an endosome. C. 5 min; note gold in vesicles near Golgi cisternae. D. 30 min; note numerous gold sitings in mutivesicular body. Bar 100 nm. (From Morris, R. E., Gerstein, A. S., Bonventre, P. F., and Saelinger, C. B., *Infect. Immun.*, 50, 721, 1985. Reprinted by permission of the American Society of Microbiology.)

organelles, which is common to the pathway taken by ligands essential for the cell's well being.

As already described, methylamine reversibly protects cells from the action of diphtheria toxin. When Vero or BS-C-1 cells are pretreated with methylamine and biotinyl-toxin internalization is followed, a picture is obtained which is strikingly different from that seen in the absence of amine.[13] In the presence of amine, toxin binds randomly to the cell surface at 4°C and the level of binding is similar to that in untreated cells. Upon warming monolayers to 37°C, toxin is cleared from the cell surface; however, it does not enter cells through coated areas of the membrane. After internalization, diphtheria toxin-gold is seen in large electron-lucent vacuoles; formation of these vacuoles is presumed to have been induced by treatment with methylamine, and thus they are believed to represent vesicles that were initially acidic. Toxin is sequestered in these vacuoles for the duration of the experiment. This suggests that methylamine is protective because it diverts toxin from the normal trafficking pathway required for toxicity. Conversely, it supports the role of coated-pit-mediated internalization as essential for toxicity.

XI. TRAFFICKING OF DIPHTHERIA TOXIN IN DIPHTHERIA TOXIN-RESISTANT CELLS

Mice and rats have long been recognized to be naturally resistant to diphtheria toxin. Bonventre and Imhoff[105] reported that diphtheria toxin inhibits protein synthesis in primary heart cell cultures from embryonic guinea pigs but not from neonatal rats. Similarly, they reported that PMN and macrophages obtained from mice are highly resistant to the action of diphtheria toxin while human and guinea pig cells are toxin sensitive.[106] Many laboratories have shown that the mouse L929 cell or its derivative, LM, is highly resistant to the action of diphtheria toxin. This resistance is not due to the inability of toxin to ADP-ribosylate elongation factor 2 from these cells; extracts of cells, tested in a cell-free, amino acid incorporation system, are as sensitive to diphtheria toxin action as are extracts prepared from sensitive cells. In addition, elongation factor 2 from rodent cells is susceptible to ADP-ribosylation by toxin. Instead, the resistance of mouse cells to diphtheria toxin has been attributed to failure of the toxin to bind to cells[4] or to nonproductive internalization of toxin by these cells.[107,108] Early work by Bonventre et al.[107] showed that iodinated diphtheria toxin is internalized by L929 fibroblasts more efficiently than by Hep-2 cells; this suggests that toxin resistance is not due to a failure of toxin to interact with cells at the cell surface. More recently, L929 cells have been shown to have surface receptors for diphtheria toxin.[34,108,109] Thus, it is generally felt that L cells are resistant to diphtheria toxin due to a defect in the handling of toxin at a step subsequent to binding.

We[108] have followed the entry of diphtheria toxin on the ultrastructural level in mouse LM cells using biotinyl-diphtheria toxin and succinyl-avidin-gold. The trafficking of diphtheria toxin in LM cells is decidedly different from that in BS-C-1 or Vero cells. Biotinyl-diphtheria toxin binds randomly to the LM cell surface and approximately 60% of this binding is displaced in the presence of native diphtheria toxin (4°C). When cells are warmed to 37°C, toxin does not move to coated areas of the cell membrane; instead, it rapidly enters cells through noncoated areas and is delivered in smooth vesicles to lysosomes. Within 5 min of warming cells to 37°C, small amounts of toxin are seen in lysosomal organelles and by 30 min, nearly 40% of all intracellular toxin is located in these vesicles. Diphtheria toxin is not seen in the Golgi region of the cell. Based on our data, it appears that diphtheria toxin follows a different pathway in LM cells than in monkey kidney cells. In addition, in LM cells, diphtheria toxin follows a pathway distinct from that taken by pseudomonas exotoxin A, a toxin which kills the LM cell in low concentrations (see Chapter 4). Keen et al.,[104] using epifluorescence video intensification microscopy, showed that diphtheria toxin and α-2 macroglobulin, a ligand known to be internalized by receptor-mediated endocytosis and delivered to lysosomes, are seen together 30 min after binding cells to 37°C. This is true in both mouse 3T3 cells (toxin resistant) and in human WI-38 cells (toxin sensitive), and can be interpreted to mean that in both cell types an easily detectable amount of toxin is ultimately delivered to lysosomes.

In an attempt to further elucidate the basis for cellular resistance to diphtheria toxin, we have tested the ability of different subcellular organelles, derived from LM cells by Percoll density radient fractionation, to convert toxin from an enzyme inactive to an enzyme active form. Gradient fractions are mixed in a test tube with toxin, then ADP-ribosyltransferase activity is measured. Preliminary experiments suggest that both endosomal-enriched and lysosomal-enriched subcellular fractions from LM cells are able to activate diphtheria toxin, if activation is carried out at an acidic pH. Thus, it is probable that LM cells are able to convert diphtheria toxin to an active form, but that the toxin either is not correctly associated with the endosomal membrane and, therefore, cannot cross the membrane to enter the cytosol, or it is delivered too rapidly to the lysosomes to have time to escape into the cytosol.

High concentrations of diphtheria toxin incubated with cells over relatively long periods of time are toxic for L cells. This toxic effect is not reduced by incubation in the presence of monensin or ammonium chloride.[111] Similarly, the resistance of LM cells to diphtheria toxin cannot be overcome by low pH. All of these facts indicate that the resistance of mouse fibroblasts to diphtheria toxin is due to a defect in toxin processing which occurs after binding but prior to translocation into the cytosol.

Conjugates of diphtheria toxin with transferrin[112] or with concanavalin A[111] also have been tested for their ability to inhibit protein synthesis in mouse L cells, with the expectation that this would lead to an understanding of the lesion in L cells which renders them highly resistant to this toxin. In both cases, diphtheria toxin is linked to the "carrier" molecule by a disulfide bridge, following introduction of a thiol group into concanavalin A or diferric transferrin with N-succinimidyl-3-2(pyridyldithio) propionate (SPDP). The transferrin-diphtheria toxin conjugate is highly toxic for LMTK$^-$ cells[112] and native transferrin totally abolishes this toxicity. Thus, the conjugate interacts with the cells via the transferrin receptors. LMTK$^-$ cells have approximately 20,000 transferrin binding sites, compared to 16,000 binding sites for diphtheria toxin.[34] Both endocytosis and exposure to an acidic environment are required for the conjugate to be toxic. In addition, exposure of conjugate-treated cells to acidic medium overcomes the protection afforded by ammonium chloride. *In toto,* the results suggest that the transferrin-toxin conjugate binds to cells via the transferrin receptor, is internalized and reaches acidic vesicles, and that fragment A enters the cell cytoplasm in response to this acidic pH.

The concanavalin A-diphtheria toxin conjugate also is highly toxic to mouse L cells.[111] Binding is via the Con A moiety, as cells are protected by treatment with α-methyl mannoside. Ammonium chloride also protects cells from the conjugate, suggesting that an acidic organelle is involved in entry into the cytosol. However, in contrast to the results with the transferrin-diphtheria toxin conjugate, exposure of L cells with prebound conjugate to a low pH does not result in toxicity. Similar treatment of Vero cells does induce toxin entry into the cytosol. In addition, the anion-transport inhibitor SITS protects Vero but not L cells from the conjugate. The authors conclude that the toxin in the Con A-diphtheria toxin conjugate enters mouse cells by a route which includes passage through a compartment of low pH, but that this route is not identical to the one followed in Vero cells.

Recently, Johnson et al.[113] conjugated CRM 107 (reduced binding and toxicity, but normal entry activity) to transferrin. This conjugate effectively killed the mouse T cell lymphoma line EL-4. Toxicity of the conjugate is 35,000-fold greater than that of native toxin alone. Their results suggest that the defect in mouse cells is at the level of the receptor (e.g., proper routing) and not in the translocation of toxin into the cytosol.

XII. FUTURE DIRECTIONS IN DIPHTHERIA TOXIN USE

Diphtheria toxin is a unique protein with potential for use in many systems not anticipated 20 years ago. Since very small amounts of fragment A can kill a cell,[41,42] it can be used in systems which require efficient cellular toxicity. The toxin has been extensively characterized both genetically and biochemically. In addition, mutant forms of toxin which have normal enzyme activity but modified binding or translocation activity have been derived. Finally, the toxin gene has been cloned and sequenced[114,115] and has been adapted for expression in mammalian cells.[116] All of these characteristics make the toxin an excellent tool for use in studying/altering mammalian cell systems. In the future, it is anticipated that new uses will be found for this potent toxic protein. Thus, what was once a significant cause of death in the human population is now being manipulated to benefit mankind.

One use of diphtheria toxin is in the formation of hybrid toxins having potential therapeutic adaptations. A second use is to ablate growth of specific cells in animals in order to investigate cell origins and development. Magic bullets or hybrid toxins composed of bacterial or plant toxins conjugate to antibodies or other carriers which specifically recognize cells are being studied for their potential therapeutic use in the treatment of disease. These conjugated toxins will be described in detail in Chapters 8 and 9 and only briefly mentioned here since diphtheria toxin has been used extensively in their design. There are several potential applications for immunotoxin therapy including treatment of cancers located in body cavities (e.g., ovarian and bladder cancers), systemic treatment of leukemia or lymphomas, use in bone marrow transplantation to destroy activated T lymphocytes involved in graft vs. host disease, and possibly in treatment of autoimmune disorders and certain parasitic diseases.[117,118]

Intact diphtheria toxin and fragment A of diphtheria toxin both have been linked to a variety of molecules. In fact, A-chains linked through disulfide bridges are now used more often then intact toxin, as this ablates the potential of the conjugate to bind, at least in part, to cell surface receptors by the B fragment. A large number of chimeras have been made using polyclonal and monoclonal antibodies. Moolten and Cooperband[119] first linked whole diphtheria toxin to the IgG fraction of antiserum to mumps virus, and showed that this hybrid was slightly more toxic to cells infected with the mumps virus than to control cells. Moolten et al.[120] attempted to treat animals with chimeras of toxin and antibody against SV-40-transformed lymphoma and sarcoma cells; some regression of tumors was achieved in mice following repeated injections of conjugates. Fragment A coupled to a monoclonal antibody directed against a colo-rectal tumor associated antigen showed selected cytotoxic effect against colo-rectal carcinoma cells in culture. However, the cytotoxic effects of the conjugate were far less than those of the native toxin.[121] Fragment A also has been linked to hormones, e.g., to EGF[122] or to insulin.[123] These conjugates are toxic only to cells having receptors for the hormone used. The study of toxic conjugates is ongoing and much needs to be learned about the optimal method of preparing the conjugates, the intracellular routing of the conjugates, and ways to overcome the immune response to the conjugates.

Maxwell and colleagues[124,125] have developed a novel use for diphtheria toxin, i.e., the investigation of the origins and interrelations of cell lineage during development. In these studies, a chimeric gene in which a cell-specific enhancer/promoter is used to drive the expression of a toxic gene product, here diphtheria toxin fragment A, is microinjected into fertilized eggs. This results in specific ablation of a specific cell lineage in the transgenic mice. To date, this method has been used to ablate normal development of cells in the pancreas[124] and in the eye lens.[125]

It is believed that every differentiated cell type expresses at least one unique gene product. Once the gene for this product is cloned and the enhancer element identified, then the targeted expression of a toxin gene should be able to be used to eliminate any cell type. Theoretically, it requires only one toxin molecule in the cell cytoplasm to kill a cell;[42] therefore, the cell would not need to synthesize a large amount of functional fragment A to achieve an effect. It should be noted that Palmiter et al.[124] calculated that it actually requires at least 300 molecules of fragment A microinjected into fertilized eggs to show any effect on development of eggs to the blastocyst stage. This requirement for a higher concentration of toxin may be explained by the fact that eggs are several hundred times larger then somatic cells. The fragment A gene used lacks a signal peptide and, therefore, should be limited to the cell cytoplasm. If released from dying cells, it would lack a B fragment and, therefore, would not enter other nearby cells. In the hybirds constructed to date, the coding region of diphtheria toxin fragment A was supplied with an elastase I enhancer/promoter to selectively direct the hybrid to pancreatic acinar cells, or with the mouse gamma 2-crystallin promoter to target the fragment A gene to the mouse eye lens. In the first instance, 7 out of 24 transgenic mice carrying the diphtheria toxin fragment A gene construct were born which lacked a pancreas; in some instances histological studies showed that these mice had a small rudiment which resembles an embryonic pancreas. In the second study, three out of six transgenic mice generated were microphthalmic; the lens of these animals displayed considerable heterogeneity ranging from reduced size to grossly aberrant morphology. All the transgenic offspring analyzed were microphthalmic indicating that this programed ablation of specific cell types can be stably transmitted through the germ line.

It should be pointed out that the gene constructs are not as toxic as had been expected. This may be explained by low efficiency of transcription or translation, or inefficient processing of transcripts. In conclusion, this methodology may eventually be applicable to elimination of any cell type including stem cells, and thus afford the potential for analyzing cellular interactions during development.

XIII. CONCLUSIONS

Diphtheria has been considered a prototype disease in that there was an early understanding of its pathogenesis at the molecular level. At the present time, the majority of research on diphtheria is directed toward the toxin and the contribution of different domains to activity. In addition, diphtheria toxin and other protein toxins are being used to increase our overall knowledge of the processing of proteins by eukaryotic cells. Major differences are seen in the way cells handle these toxic substances, from the initial binding to cells and subsequent internalization and entry into the cytosol, through their final substrate. Two toxins, such as diphtheria toxin and pseudomonas exotoxin A (Chapter 4), which have a similar end point within the cell, are believed to be trafficked in different manners. For example, acid shock of sensitive cells with diphtheria toxin bound to the surface results in translocation of toxin into the cytosol. Similarly acidic pH induces insertion of diphtheria toxin into lipid bilayer membrane systems or into liposomes. Translocation of pseudomonas exotoxin under similar conditions has been seen only with certain cell types and has not been described in artificial systems. Both toxins have been shown to enter toxin-sensitive cells through coated pits and coated vesicles; however, the organelle from which the toxin enters the cytosol may differ, with a role for the Golgi being proposed for pseudomonas exotoxin. Lastly, several of the lines of mutant CHO cells exhibit differential sensitivity to pseudomonas and diphtheria toxin. Diphtheria toxin also has been compared with ricin (Chapter 6) and again, differences in internalization and intracellular routing have been described. Reasons for these differences are unknown but are presumed to be due to differences in toxin structure, and thus in the receptor to which toxins bind.

The following model is presented for the intoxication of susceptible cells by diphtheria toxin. While alternate steps have been discussed, this model takes into account most available data.

1. Binding of toxin by the B region to a receptor on the cell surface.
2. Internalization in endosomes and subsequent exposure to an acidic environment.
3. ''Processing'' of toxin and conformational change in toxin structure to allow insertion of the toxin molecule into the organelle membrane. The receptor is generally thought

to be required to direct toxin to the proper intracellular organelle, but not for translocation to the cytosol.

4.   Release of fragment A into the cytosol, with possible reduction of the disulfide bond.
5.   Inactivation of elongation factor 2.

To date, the role of an acidic environment has not been fully delineated. It is postulated that an acidic environment is needed for insertion into the membrane, but not for entry into the cytosol. However, exposure to an acidic pH does not appear to be sufficient for penetration. While the road map leading from the cell surface to elongation factor 2 has been sketched out, many details in the routing must be filled in by further investigations.

ACKNOWLEDGMENTS

The author wishes to thank Dr. Randal E. Morris for the electron micrographs and useful discussions.

The investigations performed by the author were supported by grants from the National Institutes of Health, AI 17529 and GM 24028.

REFERENCES

1. **Loeffler, F.,** Untersuchung über die Bedeutung der Mikroorganismen für die Entstehung der Diphtherie beim Menschen, bei der Taube und beim Kalbe, *Mitt. Gesundheitsamte,* 2, 421, 1884.
2. **Roux, E. and Yersin, A.,** Contribution à l'étude de la diphthériae, *Ann. Inst. Pasteur,* 2, 629, 1888.
3. **Von Behring, E. and Kitasato, S.,** Über das Zastandenkommen der Diphtherie-immunität und der Tetanus-immunität bei Thieren, *Dtsch. Med. Wochenschr.,* 16, 1113, 1890.
4. **Pappenheimer, A. M. Jr.,** Diphtheria toxin, *Annu. Rev. Biochem.,* 46, 69, 1977.
5. **Middlebrook, J. L. and Dorland, R. B.,** Bacterial toxins: cellular mechanisms of action, *Microbiol. Rev.,* 48, 199, 1984.
6. **Neville, D. M. and Hudson, T. H.,** Transmembrane transport of diphtheria toxin, related toxins, and colicins, *Annu. Rev. Biochem.,* 55, 195, 1986.
7. **Bishai, W. R., Miyanohara, A., and Murphy, J. R.,** Cloning and expression in *Escherichia coli* of three fragments of diphtheria toxin truncated within fragment B, *J. Bacteriol.,* 169, 1554, 1987.
8. **Barbieri, J. T. and Collier, R. J.,** Expression of a mutant, full-length form of diphtheria toxin in *Escherichia coli, Infect. Immun.,* 55, 1647, 1987.
9. **Eidels, L., Proia, R. L., and Hart, D. A.,** Membrane receptors for bacterial toxins, *Microbiol. Rev.,* 47, 596, 1983.
10. **Middlebrook, J. L., Dorland, R. B., and Leppla, S. H.,** Association of diphtheria toxin with Vero cells: demonstration of a receptor, *J. Biol. Chem.,* 253, 7325, 1978.
11. **Sandvig, K. and Olsnes, S.,** Entry of the toxic proteins abrin, modeccin, ricin, and diphtheria toxin into cells: Requirement for calcium, *J. Biol. Chem.,* 257, 7495, 1982.
12. **Schaefer, E. M., Moehring, J. M., and Moehring, T. J.,** Binding of diphtheria toxin to CHO-KI and Vero cells is dependent on cell density, *J. Cell. Physiol.,* 135, 407, 1988.
13. **Morris, R. E., Gerstein, A. S., Bonventre, P. F., and Saelinger, C. B.,** Receptor-mediated entry of diphtheria toxin into monkey kidney (Vero) cells: electron microscope evaluation, *Infect. Immun.,* 50, 721, 1985.
14. **Middlebrook, J. L. and Dorland, R. B.,** Protection of mammalian cells from diphtheria toxin by exogenous nucleotides, *Can. J. Microbiol.,* 25, 285, 1979.
15. **Lory, S. and Collier, R. J.,** Diphtheria toxin: nucleotide binding and toxin heterogeneity, *Proc. Natl. Acad. Sci. U.S.A.,* 77, 267, 1980.
16. **Proia, R. L., Hart, D. A., and Eidels, L.,** Interaction of diphtheria toxin with phosphorylated molecules, *Infect. Immun.,* 26, 942, 1979.
17. **Uchida, T., Pappenheimer, A. M., Jr., and Harper, A. A.,** Diphtheria toxin and related proteins. II. Kinetic studies on intoxication of HeLa cells by diphtheria toxin and related proteins, *J. Biol. Chem.,* 248, 3845, 1973.

18. **Mekada, E. and Uchida, T.,** Binding properties of diphtheria toxin to cells are altered by mutation in the fragment A domain, *J. Biol. Chem.,* 260, 12148, 1985.

19. **Hu, V. W. and Holmes, R. K.,** Single mutation in the A domain of diphtheria toxin results in a protein with altered membrane insertion behavior, *Biochim. Biophys. Acta,* 902, 24, 1987.

20. **Papini, E., Colonna, R., Schiavo, G., Cusinato, F., Tomasi, M., Rappuoli, R., and Montecucco, C.,** Diphtheria toxin and its mutant *crm* 197 differ in their interaction with lipids, *FEBS,* 215, 73, 1987.

21. **Moehring, T. J. and Crispell, J. P.,** Enzyme treatment of KB cells: the altered effect of diphtheria toxin, *Biochem. Biophys. Res. Commun.,* 60 1446, 1974.

22. **Olsnes, S., Carvajal, E., Sundan, A., and Sandvig, K.,** Evidence that membrane phospholipids and protein are required for binding of diphtheria toxin in Vero cells, *Biochim. Biophys. Acta,* 846, 334, 1985.

23. **Olsnes, S. and Sandvig, K.,** Interaction between diphtheria toxin entry and anion transport in Vero cells, II. Inhibition of anion antiport by diphtheria toxin, *J. Biol. Chem.,* 261, 1553, 1986.

24. **Sandvig, K. and Olsnes, S.,** Interactions between diphtheria toxin entry and anion transport in Vero cells, IV. Evidence that entry of diphtheria toxin is dependent on efficient anion transport, *J. Biol. Chem.,* 261, 1570, 1986.

25. **Sandvig, K., Sundan, A., and Olsnes, S.,** Effect of potassium depletion of cells on their sensitivity to diphtheria toxin and pseudomonas toxin, *J. Cell. Physiol.,* 124, 54, 1985.

26. **Alving, C. R., Iglewski, B. H., Urban, K. A., Moss, J., Richards, R. L., and Sadoff, J. C.,** Binding of diphtheria toxin to phospholids in liposomes, *Proc. Natl. Acad. Sci. U.S.A.,* 77, 1986, 1980.

27. **Boquet, P. and Duflot, E.,** Studies on the role of a nucleoside-phosphate-binding site of diphtheria toxin in the binding of toxin to Vero cells or liposomes, *Eur. J. Biochem.,* 121, 93, 1981.

28. **Chin, D. and Simon, M. I.,** Diphtheria toxin interaction with an inhibitory activity from plasma membranes, in *Receptor Mediated Binding and Internalization of Toxins and Hormones,* Middlebrook, J. L. and Kohn, L. D., Eds., Academic Press, New York, 1981, 53.

29. **Proia, R. L., Eidels, L., and Hart, D. A.,** Diphtheria toxin-binding glycoproteins on hamster cells: candidates for diphtheria toxin receptors, *Infect. Immun.,* 25, 786, 1979.

30. **Proia, R. L., Hart, D. A., Holmes, R. K., Holmes, K. V., and Eidels, L.,** Immunoprecipitation and partial characterization of diphtheria toxin-binding glycoproteins from surface of guinea pig cells, *Proc. Natl. Acad. Sci. U.S.A.,* 76, 685, 1979.

31. **Proia, R. L., Eidels, L., and Hart, D. A.,** Diphtheria toxin: receptor interaction: characterization of the receptor interaction with the nucleotide-free toxin, the nucleotide-bound toxin, and the B-fragment of toxin, *J. Biol. Chem.,* 256, 4991, 1981.

32. **Cieplak, W., Gaudin, H. M., and Eidels, L.,** Diphtheria toxin receptor: Identification of specific diphtheria toxin-binding proteins on the surface of Vero and BS-C-1 cells, *J. Biol. Chem.,* 262, 13246, 1987.

33. **Hranitzky, K. W., Durham, D. L., Hart, D. A., and Eidels, L.,** Role of glycosylation in expression of functional diphtheria toxin receptors, *Infect. Immun.,* 49, 336, 1985.

34. **Didsbury, J. R., Moehring, J. M., and Moehring, T. J.,** Binding and uptake of diphtheria toxin by toxin-resistant Chinese hamster ovary and mouse cells, *Mol. Cell. Biol.,* 3, 1283, 1983.

35. **Sandvig, K. and Olsnes, S.,** Anion requirement and effect of anion transport inhibitors on the response of Vero cells to diphtheria toxin and modeccin, *J. Cell. Physiol.,* 119, 7, 1984.

36. **Olsnes, S., Carvajal, E., and Sandvig, K.,** Interactions between diphtheria toxin entry and anion transport in Vero cells, III. Effect on toxin binding and anion transport of tumor promoting phorbol esters, vanadate, fluoride, and salicylate, *J. Biol. Chem.,* 261, 1562, 1986.

37. **Mekada, E., Okada, Y., and Uchida, T.,** Identification of diphtheria toxin receptor and a nonproteinous diphtheria-toxin binding molecule in Vero cell membrane, *J. Cell Biol.,* 107, 511, 1988.

38. **Montecucco, C.,** How do tetanus and botulinum toxins bind to neuronal membranes?, *Trends Biochem. Sci.,* 11, 314, 1986.

39. **Creagan, R. P., Chen, S., and Ruddle, F. H.,** Genetic analysis of the cell surface: Association of human chromosome 5 with sensitivity to diphtheria toxin in mouse-human somatic cell hybrids, *Proc. Natl. Acad. Sci. U.S.A.,* 72, 2237, 1975.

40. **Hayes, H., Kaneda, Y., Uchida, T., and Okada, Y.,** Regional assignment of the gene for diphtheria toxin sensitivity using subchromosomal fragments in microcell hybrids, *Chromosoma,* 96, 26, 1987.

41. **Moynihan, M. R. and Pappenheimer, A. M., Jr.,** Kinetics of adenosinediphoshoribosylation of elongation factor 2 in cells exposed to diphtheria toxin, *Infect. Immun.,* 32, 575, 1981.

42. **Yamaizumi, M., Mekada, E., Uchida, T., and Okada, Y.,** One molecule of diphtheria toxin fragment A introduced into a cell can kill the cell, *Cell,* 15, 245, 1978.

43. **Neutra, M. R., Ciechanover, A., Owen, L. S., and Lodish, H. F.,** Intracellular transport of transferrin- and asialoorosomucoid-colloidal gold conjugates to lysosomes after receptor-mediated endocytosis, *J. Histochem. Cytochem.,* 33, 1134, 1985.

44. **van Deurs, B., Tonnessen, T. I., Petersen, O. W., Sandvig, K., and Olsnes, S.,** Routing of internalized ricin and ricin conjugates to the golgi complex, *J. Cell Biol.,* 102, 37, 1986.

45. **Marnell, M. H., Shia, S.-P., Stookey, M., and Draper, R. K.,** Evidence for penetration of diphtheria toxin to the cytosol through a prelysosomal membrane, *Infect. Immun.*, 44, 145, 1984.

46. **Larkin, J. M., Brown, M. S., Goldstein, J. L., and Anderson, R. G. W.,** Depletion of intracellular potassium arrests coated pit formation and receptor-mediated endocytosis in fibroblasts, *Cell*, 33, 273, 1983.

47. **Larkin, J. M., Donzell, W. C., and Anderson, R. G. W.,** Modulation of intracellular potassium and ATP: effects on coated pit function in fibroblasts and hepatocytes, *J. Cell. Physiol.*, 124, 372, 1985.

48. **Moya, M., Dautry-Varsat, A., Goud, B., Louvard, D., and Boquet, P.,** Inhibition of coated pit formation in Hep$_2$ cells blocks the cytotoxicity of diphtheria toxin but not that of ricin toxin, *J. Cell. Biol.*, 101, 548, 1985.

49. **Madshus, I. H., Tønnessen, T. I., Olsnes, S., and Sandvig, K.,** Effect of potassium depletion of Hep-2 cells on intracellular pH and on chloride uptake by anion antiport, *J. Cell. Physiol.*, 131, 6, 1987.

50. **Kim, K. and Groman, N. B.,** *In vitro* inhibition of diphtheria toxin action by ammonium salts and amines, *J. Bacteriol.*, 90, 1552, 1965.

51. **Kim, K. and Groman, N. B.,** Mode of inhibition of diphtheria toxin by ammonium chloride, *J. Bacteriol.*, 90, 1557, 1965.

52. **Ivins, B., Saelinger, C. B., Bonventre, P. F., and Woscinski, C.,** Chemical modulation of diphtheria toxin action on cultured mammalian cells, *Infect. Immun.*, 11, 665, 1975.

53. **Middlebrook, J. L. and Dorland, R. B.,** Differential chemical protection of mammalian cells from the exotoxins of *Corynebacterium diphtheriae* and *Pseudomonas aeruginosa*, *Infect. Immun.*, 16, 232, 1977.

54. **Dorland, R. B.,** The protective mechanism of action of amines in diphtheria toxin treated Vero cells, *Can. J. Microbiol.*, 28, 611, 1982.

55. **Dorland, R. B., Middlebrook, J. L., and Leppla, S. H.,** Effect of ammonium chloride on receptor-mediated uptake of diphtheria toxin by Vero cells, *Exp. Cell Res.*, 134, 319, 1981.

56. **Draper, R. K. and Simon, M. I.,** The entry of diphtheria toxin into the mammalian cell cytoplasm: evidence for lysosomal involvement, *J. Cell. Biol.*, 87, 849, 1980.

57. **Mekada, E., Uchida, T., and Okada, Y.,** Methylamine stimulates the action of ricin toxin but inhibits that of diphtheria toxin, *J. Biol. Chem.*, 256, 1225, 1981.

58. **Sandvig, K. and Olsnes, S.,** Diphtheria toxin entry into cells is facilitated by low pH, *J. Cell Biol.*, 87, 828, 1980.

59. **Sandvig, K. and Olsnes, S.,** Rapid entry of nicked diphtheria toxin into cells at low pH, *J. Biol. Chem.*, 256, 9068, 1981.

60. **Sandvig, K. and Olsnes, S.,** Entry of the toxic proteins abrin, modeccin, ricin and diphtheria toxin into cells, II. Effect of pH, metabolic inhibitors, and ionophores and evidence for toxin penetration from endocytic vesicles, *J. Biol. Chem.*, 257, 7504, 1982.

61. **Middlebrook, J. L.,** Effect of energy inhibitors on cell surface diphtheria toxin receptor numbers, *J. Biol. Chem.*, 256, 7898, 1981.

62. **Blewitt, M. G., Zhao, J.-M., McKeever, B., Sarma, R., and London, E.,** Fluorescence characterization of the low pH-induced change in diphtheria toxin conformation: effect of salt, *Biochem. Biophys. Res. Commun.*, 120, 286, 1984.

63. **Blewitt, M. G., Chung, L. A., and London, E.,** Effect of pH on the conformation of diphtheria toxin and its implications for membrane penetration, *Biochemistry*, 24, 5458, 1985.

64. **London, E., Zhao,, J.-M., Chattopadhyay, A., Blewitt, M., McKeever, B., and Sarma, R.,** Fluorescence quenching by a brominated detergent: application to diphtheria toxin structure, *Ann. N.Y. Acad. Sci.*, 435, 558, 1984.

65. **Dumont, M. E. and Richards, F. M.,** The pH-dependent conformational change of diphtheria toxin, *J. Biol. Chem.*, 263, 2087, 1988.

66. **Donovan, J. J., Simon, M. I., Draper, R. K., and Montal, M.,** Diphtheria toxin forms transmembrane channels in planar lipid bilayers, *Proc. Natl. Acad. Sci. U.S.A.*, 78, 172, 1981.

67. **Donovan, J. J., Simon, M. I., and Montal, M.,** Insertion of diphtheria toxin into and across membranes: role of phosphoinositide asymmetry, *Nature*, 298, 669, 1982.

68. **Kagan, B. L., Finkelstein, A., and Colombini, M.,** Diphtheria toxin fragment forms large pores in phospholipid bilayer membranes, *Proc. Natl. Acad. Sci. U.S.A.*, 78, 4950, 1981.

69. **Shiver, J. W. and Donovan, J. J.,** Interaction of diphtheria toxin with lipid vesicles: determinants of ion channel formation, *Biochim. Biophys. Acta*, 903, 48, 1987.

70. **Hoch, D. H., Romero-Mira, M., Ehrlich, B. E., Finkelstein, A., DasGupta, B. R., and Simpson, L. L.,** Channels formed by botulinum, tetanus, and diphtheria toxins in planar lipid bilayers; relevance to translocation of proteins across membranes, *Proc. Natl. Acad. Sci. U.S.A.*, 82, 1692, 1985.

71. **Sandvig, K. and Olsnes, S.,** Diphtheria toxin-induced channels in Vero cells selective for monovalent cations, *J. Biol. Chem.*, 263, 12352, 1988.

72. **Donovan, J. J., Simon, M. I., and Montal, M.,** Requirements for the translocation of diphtheria toxin fragment A across lipid membranes, *J. Biol. Chem.*, 260, 8817, 1985.

73. **Sandvig, K., Tønnessen, T. I., Sand, O., and Olsnes, S.,** Requirement of a transmembrane pH gradient for the entry of diphtheria toxin into cells at low pH, *J. Biol. Chem.,* 261, 11639, 1986.

74. **Tønnessen, T. I., Ludt, J., Sandvig, K., and Olsnes, S.,** Bicarbonate/chloride antiport in Vero cells: I. Evidence for both sodium-linked and sodium-independent exchange, *J. Cell. Physiol.,* 132, 183, 1987.

75. **Moehring, J. M. and Moehring, T. J.,** Characterization of the diphtheria toxin-resistance system in Chinese hamster ovary cells, *Som. Cell. Genet.,* 5, 453, 1979.

76. **Kohno, K., Uchida, T., Mekada, E., and Okada, Y.,** Characterization of diphtheria-toxin-resistant mutants lacking receptor function or containing nonribosylatable elongation factor 2, *Som. Cell Mol. Genet.,* 11, 421, 1985.

77. **Kaneda, Y., Uchida, T., Mekada, E., Nakanishi, M., and Okada, Y.,** Entry of diphtheria toxin into cells: possible existence of cellular factor(s) for entry of diphtheria toxin into cells was studied in somatic cell hybrids and hybrid toxins, *J. Cell Biol.,* 98, 466, 1984.

78. **Moehring, T. J. and Moehring, J. M.,** Response of cultured mammalian cells to diphtheria toxin. V. Concurrent resistance to ribonucleic acid viruses in diphtheria toxin-resistant KB cell strains, *Infect. Immun.,* 6, 493, 1972.

79. **Haigler, H. T., Wiley, H. S., Moehring, J. M., and Moehring, T. J.,** Altered degradation of epidermal growth factor in a diphtheria toxin-resistant clone of KB cells, *J. Cell. Physiol.,* 124, 322, 1985.

80. **Marnell, M. H., Mathis, L. S., Stookey, M., Shia, S.-P., Stone, D. K., and Draper, R. K.,** A Chinese hamster ovary cell mutant with a heat-sensitive, conditional-lethal defect in vacuolar function, *J. Cell. Biol.,* 99, 1907, 1984.

81. **Timchak, L. M., Kruse, F., Marnell, M. H., and Draper, R. K.,** A thermosensitive lesion in a Chinese hamster cell mutant causing differential effects on the acidification of endosomes and lysosomes, *J. Biol. Chem.,* 261, 14154, 1986.

82. **Stone, D. K., Marnell, M., Yang, Y., and Draper, R. K.,** Thermolabile proton translocating ATPase and pump activities in a clathrin-coated vesicle fraction from an acidification defective Chinese hamster cell line, *J. Biol. Chem.,* 262, 9883, 1987.

83. **Moehring, J. M. and Moehring, T. J.,** Strains of CHO-K1 cells resistant to *Pseudomonas* exotoxin A and cross-resistant to diphtheria toxin and viruses, *Infect. Immun.,* 41, 998, 1983.

84. **Merion, M., Schlesinger, P., Brooks, R. M., Moehring, J. M., Moehring, T. J., and Sly, W. S.,** Defective acidification of endosomes in Chinese hamster ovary cell mutants "cross-resistant" to toxin and viruses, *Proc. Natl. Acad. Sci. U.S.A.,* 80, 5315, 1983.

85. **Robbins, A. R., Peng, S. S., and Marshall, J. L.,** Mutant Chinese hamster ovary cells pleiotropically defective in receptor-mediated endocytosis, *J. Cell. Biol.,* 96, 1064, 1983.

86. **Klausner, R. D., van Renswoude, J., Kempf, C., Rao, K., Bateman, J. L., and Robbins, A. R.,** Failure to release iron from transferrin in a Chinese hamster ovary cell mutant pleiotropically defective in endocytosis, *J. Cell Biol.,* 98, 1098, 1984.

87. **Yamashiro, D. J. and Maxfield, F. R.,** Kinetics of endosome acidification in mutant and wild-type Chinese hamster ovary cells, *J. Cell Biol.,* 105, 2713, 1987.

88. **Yamashiro, D. J. and Maxfield, F. R.,** Acidification of morphologically distinct endosomes in mutant and wild-type Chinese hamster ovary cells, *J. Cell Biol.,* 105, 2723, 1987.

89. **Robbins, A. R., Oliver, C., Bateman, J. L., Krag, S. S., Galloway, C. J., and Mellman, I.,** A single mutation in Chinese hamster ovary cells impairs both golgi and endosomal functions, *J. Cell Biol.,* 99, 1296, 1984.

90. **Kohno, K., Hayes, H., Mekada, E., and Uchida, T.,** Mutant with diphtheria toxin receptor and acidification function but defective in entry of toxin, *Exp. Cell Res.,* 172, 54, 1987.

91. **Greenfield, L., Johnson, V. G., and Youle, R. J.,** Mutations in diphtheria toxin separate binding from entry and amplify immunotoxin selectively, *Science,* 238, 536, 1987.

92. **Laird, W. and Groman, N.,** Isolation and characterization of tox mutants of corynebacteriophage beta, *J. Virol.,* 19, 220, 1976.

93. **Colombatti, M., Greenfield, L., and Youle, R. J.,** Cloned fragment of diphtheria toxin linked to T cell-specific antibody identifies regions of B chain active in cell entry, *J. Biol. Chem.,* 261, 3030, 1986.94

94. **Hayakawa, S., Uchida, T., Mekada, E., Moynihan, M. R., and Okada, Y.,** Monoclonal antibody against diphtheria toxin: Effect on toxin binding and entry into cells, *J. Biol. Chem.,* 258, 4311, 1983.

95. **Moskaug, J. O., Sandvig, K., and Olsnes, S.,** Low pH-induced release of diphtheria toxin A-fragment in Vero cells: biochemical evidence for transfer to the cytosol, *J. Biol. Chem.,* 263, 2518, 1988.

96. **Hudson, T. H., Scharff, J., Kimak, M. A. G., and Neville, D. M., Jr.,** Energy requirements for diphtheria toxin translocation are coupled to the maintenance of a plasma membrane poteintial and a proton gradient, *J. Biol. Chem.,* 263, 4773, 1988.

97. **Hudson, T. H. and Neville, D. M., Jr.,** Quantal entry of diphtheria toxin to the cytosol, *J. Biol. Chem.,* 260, 2675, 1985.

98. **Hudson, T. H. and Neville, D. M., Jr.,** Temporal separation of protein toxin translocation from processing events, *J. Biol. Chem.,* 262, 16484, 1987.

99. **Moskaug, J. O., Sandvig, K., and Olsnes, S.,** Cell-mediated reduction of the interfragment disulfide in nicked diphtheria toxin, *J. Biol. Chem.,* 262, 10339, 1987.

100. **Wright, H. T., Marston, A. W., and Goldstein, D. J.,** A functional role for cysteine disulfides in the transmembrane transport of diphtheria toxin, *J. Biol. Chem.,* 259, 1649, 1984.

101. **Fedde, K. N. and Sly, W. S.,** Intracellular localization and degradation of diphtheria toxin, *J. Cell. Biochem.,* 37, 233, 1988.

102. **Diment, S. and Stahl, P.,** Macrophage endosomes contain proteases which degrade endocytosed protein ligands, *J. Biol. Chem.,* 260, 15311, 1985.

103. **Ciechanover, A., Finley, D., and Varshavsky, A.,** The ubiquitin-mediated proteolytic pathway and mechanisms of energy-dependent intracellular protein degradation, *J. Cell Biochem.,* 24, 27, 1984.

104. **Keen, J. H., Maxfield, F. R., Hardegree, M. C., and Habig, W. H.,** Receptor-mediated endocytosis of diphtheria toxin by cells in culture, *Proc. Natl. Acad. Sci. U.S.A.,* 79, 2912, 1982.

105. **Bonventre, P. F. and Imhoff, J. G.,** Studies on the mode of action of diphtheria toxin. II. Protein synthesis in primary heart cell cultures, *J. Exp. Med.,* 126, 1079, 1967.

106. **Saelinger, C. B., Bonventre, P. F., and Imhoff, J. G.,** Interaction of toxin of *Corynebacterium diphtheriae* with phagocytes from susceptible and resistant species, *J. Infect. Dis.,* 131, 431, 1975.

107. **Bonventre, P. F., Saelinger, C. B., Ivins, B., Woscinski, C., and Amorini, M.,** Interaction of cultured mammalian cells with ^{125}I-diphtheria toxin, *Infect. Immun.,* 11, 675, 1975.

108. **Morris, R. E. and Saelinger, C. B.,** Diphtheria toxin does not enter resistant cells by receptor-mediated endocytosis, *Infect. Immun.,* 42, 812, 1983.

109. **Heagy, W. E., and Neville, D. M., Jr.,** Kinetics of protein synthesis inactivation by diphtheria toxin in toxin-resistant L cells, *J. Biol. Chem.,* 256, 12788, 1981.

110. **Schaefer, E. M., Moehring, J. M., and Moehring, T. J.,** A lesion that prevents receptor-mediated internalization of diphtheria toxin in mutant CHO-KI cells and mouse L cells, *Infect. Immun.,* 1988.

111. **Guillemot, J. C., Sundan, A., Olsnes, S., and Sandvig, K.,** Entry of diphtheria toxin linked to con-canavalin A into primate and murine cells, *J. Cell. Physiol.,* 122, 193, 1985.

112. **O'Keefe, D. O. and Draper, R. K.,** Characterization of a transferrin-diphtheria toxin conjugate, *J. Biol. Chem.,* 260, 932, 1985.

113. **Johnson, V. G., Wilson, D., Greenfield, L., and Youle, R. J.,** The role of the diphtheria toxin receptor in cytosol translocation, *J. Biol. Chem.,* 263, 1295, 1988.

114. **Kaczorek, M., Delpeyroux, F., Chenciner, M., Streeck, R. E., Murphy, J. R.,and Tiollais, P.,** Nucleotide sequence and expression of the diphtheria *tox* 228 gene in *Escherichia coli, Science,* 221, 855, 1983.

115. **Greenfield, L., Bjorn, M. J., Horn, G., Fong, D., Buck, G. A., Collier, R. J., and Kaplan, D. A.,** Nucleotide sequence of the structural gene for diphtheria toxin carried by the corynebacteriophage β, *Proc. Natl. Acad. Sci. U.S.A.,* 80, 6853, 1983.

116. **Maxwell, I. H., Maxwell, F., and Globe, L. M.,** Regulated expression of a diphtheria toxin A chain gene transfected into human cells: A possible strategy for inducing cancer cell suicides, *Cancer Res.,* 46, 4660, 1986.

117. **Vitetta, E. S. and Uhr, J. W.,** Immunotoxins, *Annu. Rev. Immunol.,* 3, 197, 1985.

118. **Pastan, I., Willingham, M. C., and FitzGerald, D. J. P.,** Immunotoxins, *Cell,* 47, 641, 1986.

119. **Moolten, F. L. and Cooperband, S. R.,** Selective destruction of target cells by diphtheria toxin conjugated to antibody directed against antigens on the cells, *Science,* 169, 68, 1970.

120. **Moolten, F. L., Capparell, N. J., Zajdel, S. H., and Cooperband, S. R.,** Antitumor effects of antibody-diphtheria toxin conjugates. II. Immunotherapy with conjugates directed against tumor antigens induced by simian virus 40, *J. Natl. Cancer Inst.,* 55, 473, 1975.

121. **Gilliland, D. G., Steplewski, Z., Collier, R. J., Mitchell, K. F., Chang, T. H., and Koprowski, H.,** Antibody-directed cytotoxic agents: Use of monoclonal antibody to direct the action of toxin A chains to colorectal carcinoma cells, *Proc. Natl. Acad. Sci. U.S.A.,* 77, 4539, 1980.

122. **Cawley, D. B., Herschman, H. R., Gilliland, D. G., and Collier, R. J.,** Epidermal growth factor-toxin A conjugates: EGF-ricin A is a potent toxin while EGF-diphtheria fragment A is nontoxic, *Cell,* 22, 563, 1980.

123. **Miskimins, W. K., and Shimizu, N.,** Synthesis of a cytotoxic insulin cross-linked to diphtheria toxin fragment A capable of recognizing insulin receptors, *Biochem. Biophys. Res. Commun.,* 91, 143, 1979.

124. **Palmiter, R. D., Behringer, R. R., Quaife, C. J., Maxwell, F., Maxwell, I. H., and Brinster, R. L.,** Cell lineage ablation in transgenic mice by cell-specific expression of a toxin gene, *Cell,* 50, 435, 1987.

125. **Breitman, M. L., Clapoff, S., Rossant, J., Tsui, L.-C., Glode, M., Maxwell, I. A., and Bernstein, A.,** Genetic ablation: targeted expression of a toxin gene causes microphthalmia in transgenic mice, *Science,* 238, 1563, 1987.

126. **Saelinger, C. B.,** personal observations.

Chapter 4

INTERACTION BETWEEN PSEUDOMONAS EXOTOXIN A AND MOUSE LM FIBROBLAST CELLS

Randal E. Morris

TABLE OF CONTENTS

I. INTRODUCTION

Pseudomonas exotoxin A (PE) is a 66,583-dalton protein toxin secreted by *Pseudomonas aeruginosa*. PE is toxic because it catalyzes the NAD^+-dependent ADP-ribosylation of mammalian elongation factor-2 (EF-2) which results in the cessation of protein synthesis ultimately leading to cell death;[1] thus PE is an ADP-ribosyltransferase. This mechanism of toxin action is identical to that of diphtheria toxin (DT) and is given by the following formula:

$$NAD^+ + EF\text{-}2 \xrightarrow{PE} ADP\text{-ribosyl-EF-2} + nicotinamide + H^+$$

The two toxins, however, are immunologically distinct proteins secreted by two different bacteria, and produce different clinical symptomologies. Not all mammalian cells are equally sensitive to PE and DT[2], suggesting, as is the case with viruses, that the host range is the result of a cell surface receptor for the toxins. Because the substrate for PE is a cytoplasmic protein, the toxin must gain entrance into the cytosol to exert an effect. PE as secreted by the bacterium is a proenzyme; that is, it has biological activity (the ability to kill sensitive cells) but minimal enzymatic activity (the ability to ADP-ribosylate EF-2 in a test tube assay). Conversion of the proenzyme to an enzymatically active form by reduction in the presence of urea results in loss of biological activity.[3] Taken together, these observations suggest a three-step process leading to the intoxication of sensitive cells: (1) a binding event, presumably mediated by cell surface receptors; (2) a translocation event during which the toxin enters the cytoplasm and is converted to an active enzyme; and (3) the enzymatic event in which host cell EF-2 is covalently modified. This scenario of events seems to represent a paradigm for a number of bacterial toxins including the exotoxins of *Corynebacterium diphtheria* (DT) and *Vibrio cholerae* (cholera toxin), *Escherichia coli* heat-labile toxin, and the exotoxins of *Shigella dysenteriae* (shiga toxin) and *Bacillus anthracis* (anthrax toxin). These toxins all have been shown to ADP-ribosylate substrates in mammalian cells. Recent evidence suggests that certain clostridial toxins may also be members of this group.[4] In addition, these proteins toxins all have A-B structural similarity where the A fragment is the enzymatic region and the B fragment is the binding region.

II. STRUCTURE-FUNCTION RELATIONSHIP

Recent studies by several groups using X-ray crystallography and molecular genetics have elucidated several structure-function relationships for PE. Gray et al.[5] cloned the structural gene for PE into *E. coli*. Analysis of the cloned gene product revealed that PE is made up of 638 amino acids from which a 25-amino acid hydrophobic leader peptide is removed during the secretion process. Using cloned fragments, these investigators showed that the ADP-ribosyltransferase activity is associated with the carboxy terminal portion of the molecule. Allured et al.[6] crystallized PE and examined it by X-ray crystallography at a 3.0-Å resolution. Their data show that PE is composed of three distinct structural domains. Domain I which includes residues 1 to 252 (I_a) and 365 to 404 (I_b) is an antiparallel β-structure. Domain II (residues 253 to 364) is composed of six consecutive α-helices with one disulfide linking two of the helices. Domain III (residues 405 to 613) is less regular than the other two domains. Its most notable feature is an extended cleft.

Hwang et al.[7] characterized the function of each of the domains of PE by deletion analysis. Their studies show that structural domain I_a is required for binding of the toxin, the domain II is required for translocation, and that domain III and a portion of domain I_b are required for enzymatic activity. Additional evidence, also using deletion analysis, has corroborated the finding that domain I is the binding region of PE.[8] Further analysis of domain I has shown that conversion of lysine 57 to glutamate decreases the toxicity of PE for 3T3 cells 100-fold, but only reduces mouse lethality fivefold.[9] Toxin having glutamate instead of lysine in position 57 is not able to compete for the binding of wild-type toxin. Thus lysine 57 appears to be required for binding to tissue culture cells. Since mouse lethality was reduced only fivefold by this amino acid substitution, the authors hypothesize that lysine 57 is not essential in the interaction of toxin with liver, the primary target of PE *in vivo*.[10] Further studies showed that the region responsible for liver toxicity was between amino acids 225 and 252, still part of domain I.

The role of domain II in translocation has been characterized by examining its contribution to secretion from a bacterial cell.[11] When the toxin gene is cloned into *E. coli*, the gene product normally is retained in the cytoplasm.[12] Removal of domain I results in secretion of toxin into the periplasm. When domain II alone, or domains II and III are fused to the gene for a protein which is normally not secreted, the recombinant protein is now secreted into the periplasm and into the medium. The results suggest that regions of domain II are important for secretion of the toxin molecule across the bacterial cell membrane, and by analogy are required for translocation across the mammalian cell membrane during the intoxication process. How the sequence promotes secretion is not known. Carroll and Collier[13] used photoaffinity labeling with NAD (a cofactor required for the ADP-ribosylation of EF-2) to identify the active site residue of PE. Using this technique, they identified a glutamic acid residue at position 553 (located in domain III) as the active site for PE. When this glutamic acid was substituted with aspartic acid (a conservative substitution), the ADP-ribosyltransferase activity was reduced 1,800-fold and the toxicity of the mutagenized protein was reduced by 4 logs when assayed on sensitive LM mouse fibroblasts.[12] Interestingly, a glutamic acid residue in position 148 of DT is the active site of the enzymatic domain of this protein toxin.[14] Recent studies[15] have shown a region of local amino acid homology in the active site of these two toxins. Therefore, the A region of PE corresponds to domains I_b and III, and the B region corresponds to domains I_a and II.

III. RECEPTOR-MEDIATED ENDOCYTOSIS

On a structural level, we are able to assign particular functions to discrete regions of the PE molecule. The interaction of PE with mammalian cells is less well understood. Although we, as well as others, have suggested a three-step model leading to intoxication of sensitive cells, the cellular processes involved are only beginning to be elucidated. The revelation of these events parallels our understanding of basic cell biology, particularly as it relates to how cells interact with physiological ligands, i.e., hormones and growth factors. Since *a priori* there is no reason to suspect that a cell would prossess a receptor for a toxic substance, it is our hypothesis that PE binds to cells via an "opportunistic" receptor; that is, PE structurally resembles some physiological ligand. It is the binding which dictates host-range specificity and sensitivity to the toxin. Once internalized, PE is shuttled by one of the established intracellular trafficking patterns. During its residence in the various intracellular compartments, PE is converted to the active form and escapes into the cytoplasm. The initial event leading to the expression of toxicity is the receptor-mediated binding which facilitates the entry of the toxin via receptor-mediated endocytosis (RME).

Since the classical description of RME by Brown, Goldstein and colleagues in the mid-1970s,[16,17] there has been intense interest in the internalization and intracellular routing of various ligands. It is now widely accepted that most growth factors, hormones, and transport proteins enter mammalian cells by RME. The initial event in this process is the binding of the ligand to receptors located on the plasma membrane. In the majority of the well-studied systems, the receptors are randomly distributed on the cell surface. Following receptor binding, the receptor-ligand complex is seen in thickened regions on the plasma membrane. These thickened regions proceed to invaginate forming "coated pits". The protein clathrin has been identified as that which imparts the coated appearance to the internalization vesicle.[18] During the final stages of the internalization process, the clathrin-coated pit seals and is pinched off to become a coated vesicle. The coated vesicle fuses with a noncoated vesicle, the endosome. In this manner, the receptor-ligand complex is transferred from the cell surface to a cytoplasmic compartment.

Endosomes are pleotrophic organelles which are found in all regions of the cytoplasm. Often they are vacuolar structures associated with tubular elements.[19] They may also resemble multivesicular bodies (MVB).[20-21] To date, no specific proteins or unique biochemical feature has been described for endosomes. Endosomes are defined as any uncoated vesicle which labels with endocytic tracers prior to their appearance in lysosomes.[22] The role of the endosome is to direct the internalized ligand to its proper destination. Although the mechanism directing the route taken by the endosome is unknown, it is probably related to the interaction of the receptor-ligand complex in the endosomal environment. Because endosomal membranes have ATP-dependent proton pumps,[23,24] their intraorganellar pH is between 5 and 6.[25] The acidic pH has two major effects upon internalized ligand: (1) the stability of the ligands interaction with its receptor, and (2) a conformational change associated with the internalized ligand in an acid environment.

From the studies of various ligands internalized by RME, four separate intracellular routing patterns have emerged. All four routes follow RME via clathrin-coated pits as described above. In the first group, exemplified by low density lipoprotein,[26] the ligand and receptor dissociate in the endosomal compartment owing to the acidic pH of the organelle. Subsequently, the ligand (now free in the lumen of the endosome) is routed to the lysosomal compartment for degradation while the receptor is recycled to the cell surface. This process has been elegantly demonstrated with a second ligand, asialoglycoprotein, which enters hepatocytes in an analogous manner.[27] Using double-label immunoelectron microscopy on ultrathin cryosections of rat liver in conjunction with monospecific antibodies and gold-protein A complexes, they showed co-localization of asialoglycoprotein and its receptor in coated vesicles near the plasma membrane. As the endocytic process progressed with increased time of warming, the ligand was seen within the lumen of endosomes, while the receptor was seen in tubular extensions emanating from the endosomes. Because this observation suggested dissociation and segregation of asialoglycoprotein and its receptor, they termed this compartment CURL (compartment of *u*ncoupling of *r*eceptor and *l*igand). They have used this technique to show a similar routing pattern for mannose-6-phosphate and its receptor.[28]

In the second group, the receptor-ligand complex remains associated within the endosome and the complex is recycled to the cell surface. The best example of this interaction is the iron-binding protein, transferrin.[29] A single molecule of apo-transferrin is capable to binding two Fe^{3+} ions, resulting in diferric transferrin or transferrin. Transferrin binds to cell-surface receptors which internalize the iron-binding protein by RME. In a series of very clever experiments, Dautry-Varsat et al.[30] showed that in an acidic environment (pH $\leq$ 5.5) iron dissociates from the transferrin. They further demonstrated that the transferrin receptor avidly binds apo-transferrin at acidic pH, thereby allowing the ligand to remain attached to its receptor in the endosome. Upon routing to the cell surface, the apo-transferrin is released because of the low affinity of the receptor for transferrin at neutral pH. After binding more Fe^{3+}, the cycle starts anew. These studies clearly demonstrate the acidic nature of the endosomal compartment. Taken together with the first routing mechanism (low density lipoprotein and asialoglycoprotein), the importance of the acidic environment in the intracellular trafficking of internalized ligands is established.

In the third group, both the ligand and its receptor are routed to the lysosomal compartment for degradation. The best studied ligands in this category include epidermal growth factor (EGF),[31,32] insulin,[33,34] and altered immunoglobulins, e.g., internalized by Fc receptors.[35] Other investigators have reported that a significant fraction of the EGF is routed to the Golgi region during its transit to the lysosomal compartment.[36,37] One curious feature of this intracellular trafficking pattern is the observation by DiPaola and Maxfield[38] that EGF and its receptor dissociate from one another at acidic pHs. This implies that something in addition to receptor-ligand interaction dictates the intracellular routing. Also, in some cells, a small but significant proportion of EGF receptors do recycle.[34,39] Perhaps, as suggested by Beguinot et al.,[40] this pathway is used to regulate receptor expression (i.e., down-regulation).

In the fourth group, the ligand and its receptor are transported from one cell surface to another, e.g., from the basolateral surface to the apical surface. The best characterized example in this group is secretory IgA. In the liver, dimeric IgA binds to newly synthesized receptor on the sinusoidal surface of the hepatocyte, is internalized, transported to the canalicular surface, and discharged into the bile.[41] The transit, because it is across the cell, is termed transcytosis. Geuze et al.[42] have used double-label immunoelectron microscopy to elucidate the transit. By way of camparison, they followed the intracellular routing of receptors for asialoglycoprotein, mannose-6-phosphate, and polymeric IgA. They showed that receptors for the former two ligands colocalized within the tubules of the CURL while IgA receptor showed dramatic microheterogeneity. They use this as evidence to support their hypothesis that in addition to its role in uncoupling and sorting recycling receptor from ligand, CURL serves as a compartment to segregate recycling receptors from the receptors involved in transcytosis. During its transit, the IgA receptor is proteolytically cleaved in such a manner that a portion of the receptor remains with the dimeric IgA. This receptor fragment is the secretory piece.[43,44] Since transit is a function of the endosomal compartment, this observation suggests that the endosomes may contain proteases. Others also have shown protease activity associated with endosomes.[45,46]

IV. PSEUDOMONAS EXOTOXIN A-LM CELL INTERACTION

This system of delivery of external ligands into mammalian cells is so efficient that it has been subverted by viruses as a mechanism for entry. As more information becomes available, it is apparent that bacterial toxins also take advantage of "opportunistic" binding (perhaps "mimicry" would be a more appropriate term) to enter sensitive cells. In this review, we summarize our knowledge of the events leading to inhibition of protein synthesis in mammalian cells by pseudomonas exotoxin A. As a model system, my colleagues and I have used mouse LM fibroblasts. We chose these cells because of their exquisite sensitivity to PE. I approach this review from the perspective of a cell biologist and discuss data concerning the interaction between PE and LM cells in context with what is known for other well-characterized ligands. In so doing, I propose a model to explain the means by which PE gains entrance into the cytosol. I do not discuss any further the mechanism by which EF-2 is ADP-ribosylated as this has already been well characterized and reviewed.[1,47]

A. THE BINDING EVENT

In 1980, using an unlabeled antibody technique in conjunction with horse spleen ferritin as the electron-dense marker, we followed the binding and internalization of PE by mouse LM fibroblasts.[48] We showed that PE binds to the plasma membrane of LM cells and enters via clathrin-coated pits. This was the first demonstration of entry of PE into a sensitive cell line by RME and suggested the existence of a cell surface receptor on LM cells for PE.

We next proceeded to characterize the binding by classical binding techniques using ^{125}I-labeled PE. Numerous attempts to label the toxin by conventional methods resulted in loss of its biological activity. We were eventually successful in obtaining ^{125}I-PE by using the Bolton-Hunter reagent (*N*-succinimidyl 3,4-hydroxy, 5-[^{125}I] iodophenyl propionate[49]).

Using this radiolabeled reagent, we characterized the binding of PE by LM cells. Unfortunately ^{125}I-PE binds to cells with very high nonspecificity at 4°C. Similar observations have been reported by Moehring and Moehring.[50] In an attempt to circumvent this problem, we measured binding of ^{125}I-PE on paraformaldehyde-fixed LM cells, and approach which has been used by others to quantitate receptor binding.[51,52] Using this system, we were able to show specific binding. Saturation is achieved at 5.4 nM toxin, which corresponds to about 100,000 receptors per LM cell. The saturation data are in good agreement with the minimal concentration of PE required for expression of toxicity, e.g., 1 to 2 nM. Further, binding is reversible since 50% of the bound ^{125}I-PE can be displaced by unlabeled PE during a 3-h incubation period.

We also have used a morphological technique to demonstrate PE binding to LM cells. In this technique, PE is biotinylated (B-PE) by incubation with biotinyl-N-hydroxysuccinimide ester at pH 9.2.[53] B-PE is allowed to bind to LM cells for 30 min at 4°C. After several washes, streptavidin-gold colloids,[54] average diameter of 5 nm, are added and monolayers incubated at 4°C for 30 min. Samples are subsequently processed for electron microscopy. Corroborating both the biochemical and previous morphological data, we observed saturation at 1.66 nM. Further, we are able to show specific binding by a competition assay in which B-PE is incubated with LM cells in the presence of a 200-fold excess of native toxin. Incubation in the presence of excess native toxin reduced the amount of gold on the cell surface to that seen in the absence of any toxin.[54] Calcium ions are not required for optimal binding, as judged by electron microscopy.[55]

The receptor on sensitive cells which is "usurped" by PE has not been identified. Trypsin[56] (unpublished observation) or elastase[57] treatment reversibly protects cells from toxin; sensitivity returns to normal after 4-h incubation in fresh medium. We have recently isolated a high molecular weight component from LM cells which specifically binds PE. Binding is reduced by treatment with pronase or trypsin, suggesting that at least part of the binding components is proteinaceous. The relationship of this material to the putative toxin receptor remains to be determined.[56]

B. THE TRANSLOCATION EVENT

During the translocation event, receptor-bound PE is transferred from the cell surface into the cell cytosol. This process can be subdivided into four steps: internalization, intracellular trafficking, conversion of the proenzyme to the enzymatically active form, and escape. These steps are reviewed as they concern the interaction of PE with mouse LM fibroblasts.

1. Internalization

As described earlier, data obtained using ferritin as an electron-dense marker in an unlabeled antibody technique suggested that PE entered LM cells by RME via clathrin-coated pits.[48] This observation was confirmed and further elucidated by studies using B-PE in conjunction with avidin-gold colloids. In these latter studies, B-PE was allowed to bind to LM cells for 30 min at 4°C. Following three washes with buffer to remove unbound toxin, avidin-gold colloids were added and incubated at 4°C for 30 min. Samples next were incubated at 37°C for 0.5 to 120 min and then prepared for electron microscopy. Two significant observations were made. First, B-PE was cleared from the cell surface very rapidly; the number of membrane-bound sitings was reduced to 50% in $2^{1}/_{2}$ min at 37°C. Clearance from the cell surface requires the presence of extracellular calcium.[55] Second, by 30 s at 37°C approximately one third of all gold sitings were seen within clathrin-coated pits[58] (Figure 1a). The number of sitings seen within coated pits after 1 min dropped, returning to background levels by 5 min. These results show that PE is bound and rapidly internalized by RME via clathrin-coated pits.

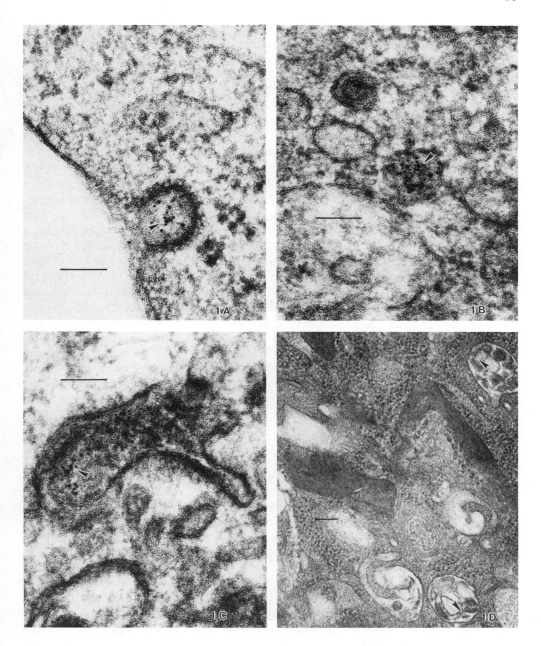

FIGURE 1.   Internalization and intracellular routing of PE by mouse LM fibroblasts. LM cell monolayers were incubated sequentially at 4°C with biotinyl-PE (100 ng/ml), washed, avidin-gold colloids added, washed, and warmed to 37°C. (A) 30 s after warming; note the PE-gold (arrowheads) in a clathrin-coated pit. (B) 1 min after warming; note PE-gold within noncoated, membrane-bound vesicle; i.e., endosome. (C) 5 min after warming; note PE-gold within a later stage endosome having tubular extensions. (D) 20 min after warming; note PE-gold within multivesicular bodies. Bar = 100 nm. (Figures 1A, B, and C are reproduced from Morris, R. E. and Saelinger, C. B., *J. Histochem. Cytochem.*, 32, 124, 1984. With permission.)

2. Intracellular Trafficking

We also have used the B-PE:avidin-gold technique to visually monitor the intracellular trafficking of the toxin in LM cells.[58-62] Shortly after entry (30 s to 2.5 min), B-PE is seen within coated vesicles. From the coated vesicles, it is transferred to noncoated, pleomorphic organelles (1 to 10 min; Figures 1b and 1c). These organelles are endosomes (see later) and route the toxin to the Golgi region. It requires about 10 to 20 min at 37°C for the toxin to move from the cell surface to the Golgi region (Figure 2b). Frequently the toxin-gold conjugates are observed in MVB (Figure 1d). Between 30 to 60 min at 37°C, the toxin begins to accumulate in the lysosomal compartment[54] (Figure 2c). We have been able to clearly identify this latter compartment as lysosomal by preloading the cells with the marker, horseradish peroxidase.[63,64]

Other investigators have shown that ligands bound to gold colloids are aberrantly routed.[65-67] However, we feel our observations using the B-PE:avidin-gold technique are valid for the following reasons. First, when LM cells are incubated with biotinyl diphtheria toxin (B-DT) and the intracellular routing of the DT-gold is then followed as described above, a different routing pattern emerges. Keep in mind that LM cells are resistant to DT[2] in spite of the fact that they have cell-surface receptors for the toxin.[68] Also DT, like PE, is an ADP-ribosyltransferase,[1] and EF-2 isolated from mouse cells is sensitive to ADP-ribosylation catalyzed by DT.[69] In contrast to the internalization and routing noted for PE-gold, DT-gold does not enter LM cells via clathrin-coated pits and is not routed to the Golgi region.[70] Second, data obtained by biochemical assays using isopycnic ultracentrifugation in conjunction with lysosomal and Golgi membrane markers (e.g., β-galactosidase, β-glucuronidase, and galactosyltransferase) corroborate the morphological observations.[59] On the basis of these findings, we suggest that entry by RME via clathrin-coated pits and routing to the Golgi region are obligatory steps for the expression of toxicity. A similar internalization and routing pattern has been shown for DT in the highly sensitive Vero cell line[70] (see Chapter 3).

3. Activation and Escape

Although the use of the biotinyl-toxin:avidin-gold method has permitted us to visualize the intracellular routing of the toxin, the site of toxin escape remains obscure. This results because the gold marker, although only 5 nm in diameter, is too large to escape with the toxin into the cytoplasm. Neither do the morphological experiments identify the site of conversion of protoxin to active toxin. To address these questions, we have used inhibitors which block toxicity, e.g., acidotropic agents and reduced temperatures.[54,58]

a. Role of the Acidic Compartments

Acidotropic agents are small lipophilic compounds which have an affinity for acidic intracellular compartments, e.g., endosomes and lysosomes. After entry into an acidic compartment, the acidotropic agent becomes protonated and unable to exit the organelle. The consequence of this is that the intraorganellar pH increases.[25,70] The net effect of the increased pH is the inhibition of intracellular routing; this implies that acidification plays a prominent role in the routing of internalized ligands.[23]

We have tested several acidotropic agents in the LM/PE cell system, including methylamine, ammonium chloride, chloroquine, and monensin.[48,60,71,72] Although there are some subtle differences in the response of the LM cell to the various agents, all four protect the cell against the toxic effects of PE. Electron microscopic studies using the B-PE:avidin-gold technique show that the acidotropic agents do not inhibit the entry of PE into LM cells. Entry in the presence of chloroquine (a tertiary amine) and monensin (a carboxylic ionophore) proceeds as described previously; entry in the presence of the primary amines, methylamine and ammonium chloride, is not mediated via clathrin-coated pits.[72] With all four agents,

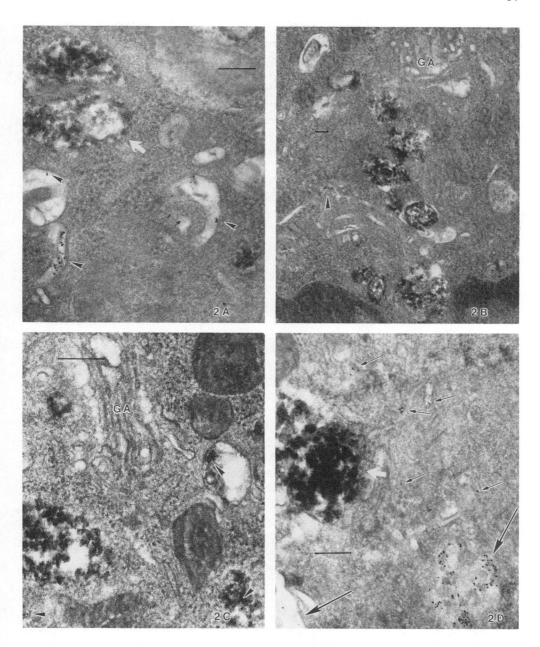

FIGURE 2. Double-labeling experiment following the intracellular trafficking of PE in mouse LM fibroblasts. LM cells were incubated with horseradish peroxidase (HRP) (1 mg/ml) for 18 h at 37°C, washed, and reincubated for 2 h at 37°C prior to the initiation of the experiment, as described in Figure 1. Amorphous black precipitate indicates the location of HRP which is used to identify the lysosomal compartment. (A) 5 min after warming to 37°C; note the PE-gold (arrowheads) within various intracellular vesicles. PE-gold is not seen within HRP-positive vesicles (white arrow). (B) 10 min after warming to 37°C; note the PE-gold within tubular elements in the perinuclear region in close proximity of the Golgi apparatus (GA). PE-gold is not seen in HRP-positive vesicles. (C) 30 min after warming to 37°C; note PE-gold in a small tubular vesicle (arrowhead) and in HRP-positive vesicles in the Golgi region (GA) of the cell. (D) 90 min after warming at 19°C; note PE-gold in a multivesicular body (large arrows) and tubular elements (small arrows) within the cytoplasm. PE-gold is not seen within the HRP-positive lysosome (white arrow). Bars = 100 nm. (Figures 2B and D are reproduced by permission from Morris, R. E. and Saelinger, C. B., *Infect. Immun.*, 52, 445, 1986.)

the level of protection is inversely related to the time of addition of the acidotropic agent following the internalization of the toxin[72,73] and the protection afforded is reversible.[56] This suggests two levels of protection: (1) inhibition of entry via clathrin-coated pits which might direct the toxin to an inappropriate intracellular compartment (the primary amines), and (2) elevation of the endosomal pH which blocks normal receptor-ligand interactions and results in inhibition of normal trafficking (all four agents).

Since the majority of our work has been done with methylamine, the rest of this discussion is restricted to this acidotropic agent. After entry in the presence of methylamine, toxin-gold conjugates accumulate along the membranes of numerous pleomorphic organelles (Figure 3b). As time of internalization increases in the presence of the agent at 37°C, more intracellular membrane-associated toxin-gold conjugates are noted and the organelles begin to become swollen. If the incubation period is allowed to go as long as 30 min, the organelles become very distended and the cell takes on a "swiss-cheese" appearance (Figure 3a). The cells at this point are still protected. If the methylamine block is released, the cells quickly lose their vacuolated appearance and again become sensitive to the toxin. Recent data suggest that after inhibitor removal, PE is returned to the cell surface, reinternalized by the normal mechanism (i.e., RME via clathrin-coated pits), and is routed in the typical manner (endosome to Golgi) which results in cell death.[131]

What is the role of the acidic environment in the expression of toxicity by PE? For physiological ligands, such as hormones and growth factors, the acid environment of endosomes influences ligand and receptor interaction which in turn influences the intracellular routing of the ligands.[23,27,28,65] Others have shown that increasing the pH acidic compartments inhibits receptor recycling.[51,74-77] Thus, proper intracellular trafficking for a given ligand is dependent upon various acidic compartments; in the absence of an acidic environment (as is induced by acidotropic agents), a given ligand is internalized but further processing is inhibited.[23]

All cells are not equally sensitive to PE; for example, BHK and BSC-1 cells are 100 times more resistant than mouse LM cells.[61,78] However, in general terms, acidotropic agents afford protection to all cell types. It is possible to increase the susceptibility of these less sensitive cells by treatment with calmodulin antagonists, such as trifluoperazine and dansylcadaverine. Similar treatments have no effect on highly sensitive cells.[78,79] Treatment with amines and ionophores which protect LM cells from PE reverses this increased sensitivity. Thus these drugs may be able to induce an efficient entry mechanism or redirect the toxin into an intracellular pathway which allows efficient expression of toxicity. Morphological studies need to be done to determine how trafficking of the toxin is reprogramed.

At this time it is instructive to consider several studies examining the effect of pH on the translocation of other bacterial toxins and viruses. Incubation of DT in an acidic pH (pH 5.0 to 5.5) causes a molecular conformational charge resulting in the exposure of a hydrophobic domain in the binding fragment of the toxin.[80-84] Several groups working with artificial planar lipid bilayers, have shown that it is a hydrophobic region of the binding fragment which inserts into and forms a channel through which the enzymatic fragment enters the cytosol. This process is pH dependent (pH <5.0) and occurs in the absence of receptor proteins.[85-87] More recently, Donovan et al.[88] incubated lipid vesicles containing EF-2 and radiolabeled NAD^+ with DT and measured the degree of ADP-ribosylation of EF-2 as a function of pH. They showed that translocation was pH dependent, being maximal at pH 4.5, and required the entire toxin molecule.

Hoch et al.[89] reported that botulinum neurotoxin B and tetanus toxin both generated pores in a lipid bilayer at acidic pH in a manner analogous to DT. Pore formation for all three toxins is maximal when the pH on the cis (protein) side is 4.0 and the trans (opposite) side is 7.0. Donovan and Middlebrook[90] showed that botulinum neurotoxin C also formed pores maximally in lipid bilayers at pH 6.1. Simpson[91] has shown that ammonium chloride

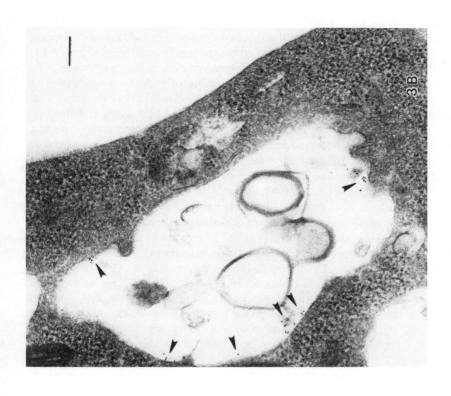

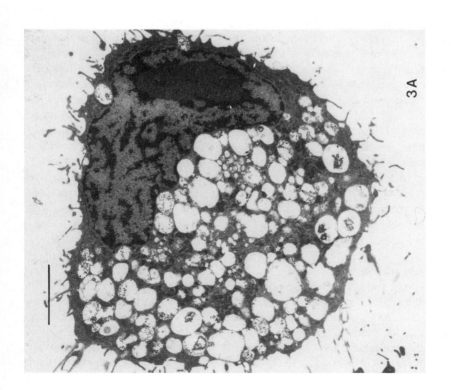

FIGURE 3. Internalization of PE by mouse LM cells in the presence of methylamine (MA). LM cells were incubated with 20 *mM* MA for 20 min at 37°C, washed with cold Hank's balanced salt solution (HBSS) containing 20 mM MA, and incubated at 4°C for 10 min. Biotinyl-PE (100 ng/ml) was added in cold HBSS containing 20 *mM* MA and incubated at 4°C for 30 min. Samples were washed, incubated with avidin–gold colloids for 30 min at 4°C, washed again, and warmed for 30 min at 37°C in HBSS containing 20 mM MA. (A) Typical "Swiss-cheese" appearance of an LM cell; bar, 1μ. (B) Higher magnification view of a swollen vesicle containing PE-gold (arrowheads). Note that the PE-gold is either on the vesicular-membrane or is attached to intravesicular materials; bar = 100 nm.

and methylamine antagonize the toxic effects of clostridial neurotoxins, again demonstrating the importance of an acidic environment.

Olsnes and colleagues have shown that the successful entry of poliovirus into sensitive cells has an acid-dependent step. They have followed the entry of poliovirus into HeLa cells. Their data show that compounds which dissipate the proton gradient across membranes, and compounds which inhibit acidification do not inhibit the binding of poliovirus, but do prevent infection.[92] They go on to show that at pH $\simeq 5.5$, the poliovirus capsid undergoes a structural alteration exposing a hydrophobic region of its icosahedral shell which leads to the escape of the viral RNA into the cytoplasm. They suggest that this conformational change is a prerequisite for infection.[93] These observations are corroborated by the electron microscopic data of Zeichhardt et al.[94] who followed the infection of Hep-2 cells by poliovirus. They show entry via clathrin-coated pits and movement to an acidic compartment. Infection, but not entry of the virus, is blocked by acidotropic agents.

FitzGerald et al.[95,96] have shown that adenovirus enters KB cells by RME. They showed that the intact virion is released into the cytoplasm of the infected cell by disruption of the endosomal membrane, and that this disruption is inhibited by various acidotropic agents.[97] Thus, the disruption of the endosomal membrane requires an acid pH. Seth et al.[98] further characterized this low pH dependency by measuring the interaction of adenovirus capsid proteins with the detergent Triton X-114 as a function of pH. The penton base protein strongly binds Triton X-114 at pH 5.5 and below, but binds the detergent weakly at pH 7.0 and above. These results suggest the expression of a hydrophobic domain in an acidic environment. Further, treatment of adenovirus with antibody to the penton base protein renders the virus noninfectious. From these data, it is inferred that the penton base protein is required for endosomal disruption.[98]

Unlike the toxins and viruses discussed above, PE does not have any long continuous stretches of hydrophobic residues. Studies by Wisnieski and colleagues have, however, revealed structural changes in the PE molecule as a function of pH.[99-101] Using an intra-membraneous photoreactive probe and monitoring the interaction of PE with liposomes of defined lipid composition, they showed that both insertion of PE into the liposomes and pore formation (i.e., the mechanism of translocation) increases with decreasing pH.[99,100] Similarly, Olsnes and Sandvig[102] showed that PE enters a detergent-rich phase at low pH, again indicating an increase in hydrophobic properties as pH decreases. More recently, Farahbakhsh et al.[101] showed that PE, like DT, undergoes a conformational change as the pH is reduced to 4.0, and that the conformational change is indicative of the exposure of hydrophobic surfaces. Although not postulated by the authors, we speculate that this would correspond to domain II (the six alpha helices) of the PE molecule. This pH-dependent conformational change is what would be expected to occur in the endosomal compartment.[91]

Taken together, these results suggest that escape from an acidic compartment (i.e., the endosomal compartment) is a common mechanism shared by several bacterial toxins and viruses. The role of the acidic environment is to cause a conformational change resulting in the expression of hydrophobic domains which can intercalate into the endosomal membrane. This permits the escape of the toxic fragment or viral nucleic acid.

b. Temperature Dependence

Normal intracellular trafficking is also influenced by temperature. Weigel and Oka[103] reported that endocytosis of [^{3}H]-asialo-orosomucoid is blocked at temperatures less than 10°C. Between 10 and 20°C, the rate of endocytosis is directly proportional to the change in temperature. At 20°C, there is a sharp increase in the rate of internalization. Dunn et al.[104] have shown that the uptake and catabolism of ^{125}I-asialofetuin increases progressively as the temperature increases from 20 to 35°C. These investigators showed that at temperatures less than 20°C, lysosomal degradation is blocked. Using both morphological and biochemical assays, Dunn et al.[104] showed that endosome-lysosome fusion is greatly reduced. Sleight and Pagano[105] studied the effect of temperature on the internalization of a fluorescenated derivative of phosphatidylcholine by hamster lung fibroblasts. They showed that at temperatures less than 12°C, the fluorescenated ligand does not enter the cells. Between 12 and 18°C, the ligand enters the cell but does not accumulate in the Golgi apparatus. Between 18 to 20°C, the fluorescent lipid enters and accumulates in the Golgi apparatus. These results suggest that the phase transition temperature, i.e., the temperature at which membrane lipids become immobile, for the lysosomal membranes is 20°C and is $\simeq$18°C for Golgi membranes.

We have measured the toxicity of PE in mouse LM fibroblasts as a function of temperature. We have shown a direct relationship between temperature and the toxicity of PE for LM cells. The TCD_{50} for PE at 37, 19, and 15°C is 0.6, 9.0, and >10,000 ng, respectively. The reduction in toxicity is not due to a reduction in ADP-ribosylation activity as the *in vitro* enzyme activity is maximal between 16 and 30°C.[49]

We have followed the intracellular routing of PE by LM cells at 19°C and 15°C at the ultrastructural level.[54,62] In these studies, we preloaded the lysosomes with horseradish peroxidase.[64] This permitted us to identify the lysosomal compartments. At 37°C, PE rapidly enters LM cells via clathrin-coated pits (30 s), is transferred to coated vesicles (1 min), is routed to the Golgi region in endosomes (5 to 20 min; Figures 2a and b) and ultimately is seen within the lysosomal compartment ($\leq$ 30 min; Fig 2c). Residence in the lysosomal compartment is characterized by the coincident siting of B-PE:avidin-gold complexes and horseradish peroxidase precipitate in a single vesicle. At 19°C, PE enters LM cells but at reduced rate. PE remains in coated pits for up to 20 min at 19°C before being internalized. In agreement with others, we were unable to observe any significant accumulation of PE in the lysosomal compartment at temperatures less than 19°C; the toxin remains in the endosomes or Golgi-associated vesicles for the duration of the experiment, i.e., 3 hr[54] (Figure 2d). At 15°C, PE remains associated with coated pits for up to 60 min. With continued incubation at 15°C, toxin is internalized but is sequestered in prelysosomal compartments.[56]

To characterize the temporal relationship of the several events leading to toxicity, we used various antagonists of toxicity together. The goal of these experiments was to determine how long after warming to 37°C was treatment with various antagonists able to protect 100% of the LM cells. PE (at 1 μg/ml) was allowed to bind for 1 h at 4°C prior to the initiation of the experiment. Cells next were warmed to 37°C. At various times after warming, the ability of the antagonists to protect the cells was evaluated. The antagonists used included (1) trypsin-pronase treatment, (2) antibody, (3) methylamine (20 m*M*), and (4) incubation at 19°C. The results (Figure 4) show that trypsin-pronase and antibody are able to protect for only seconds after warming. These treatments, which are used to inactivate surface-bound toxin, suggest that PE enters the LM cells very rapidly and efficiently. Methylamine could be added 5 to 7 min after PE entry and still fully protect the cells. This defines the acid-dependent step and correlates well with morphological evidence characterizing the intracellular location of PE in the endosomal compartment. It is possible to delay shifting cells to 19°C for 14 to 15 min after PE internalization has begun and still achieve protection. This establishes the temperature-dependent step and correlates with the time PE is seen in the Golgi region. These results suggest the following series of events leading to PE escape into the LM cell cytoplasm: (1) binding and entry via RME, (2) acidification and conformational charge, and (3) escape. The results do not support the hypothesis that acidification, activation (conversion of the protoxin to the toxin), and escape occur concurrently because PE resides in an acidic, prelysosomal compartment (methylamine-sensitive step) for 7 to 10 min prior to expression of toxicity (cold-sensitive step).

A similar observation has been reported for intoxication of Vero cells by DT. Marnell et al.[106] showed that DT encounters a prelysosomal, low-pH vesicle (i.e., endosome) by 4 min after internalization. Inhibition of protein synthesis, however, was not detected until

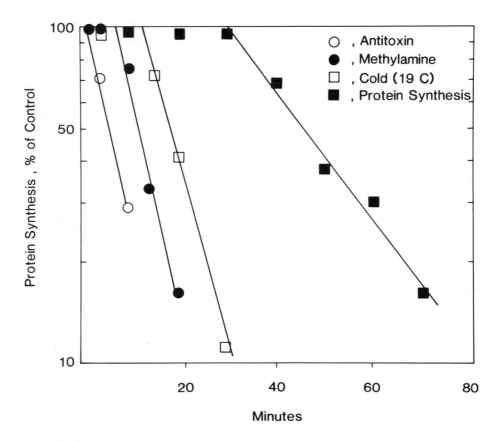

FIGURE 4.   Entry of PE. In all cases, LM monolayers were preincubated with 1 μg of PE per ml for 1 h at 4°C. (○) Cell culture fluid at 37°C was added and, at indicated times, aspirated, and monolayers were incubated for 30 min at 4°C in medium containing specific antitoxin to neutralize surface PE. Cells were then washed and incubated at 37°C for 4.5 h, prior to measurement of protein synthesis. Protein synthesis is relative to controls that were treated with antitoxin immediately after 4°C incubation with toxin. (●) Cell culture fluid at 37°C was added; when indicated, medium was replaced with 20 m*M* methylamine, and cells were incubated at 37°C for 4.5 h until protein synthesis was assayed. Protein synthesis is relative to controls which received 20 m*M* methylamine 15 min before the cells were put in 37°C medium. (□) Medium at 37°C was added, and when indicated, monolayers were transferred to 19°C medium and incubated for 4.5 h before protein synthesis was measured. Protein synthesis is relative to controls that were placed at 19°C immediately after the preincubation step. (■) Medium at 37°C was added; when indicated, 10 μCi or [³H]leucine per ml was added for 10 min. Data are relative to controls receiving no toxin and are plotted at a time 5 min after addition of label. (Reprinted by permission from Morris, R. E. and Saelinger, C. B., *Infect. Immun.*, 52, 445, 1986.)

20 min after internalization. In a detailed kinetic analysis of intoxication of Vero cells by DT, Hudson and Neville[107] suggest that intracellular acidification affects at least two separate stages of the intoxication process. This first stage would correspond to conformational changes associated with low pH. This is acid dependent and would convert the toxin to the proper conformation for pore formation. The second stage is the translocation event. Although the actual translocation is not acid dependent, it cannot occur unless the toxin is in the proper conformation (which is acid dependent). The implication of our work and that of others is that it requires about 10 min in an acidic environment for the toxins to acquire the proper conformational change necessary for escape.

The identity of these prelysosomal compartments is not known with certainty. The possibilities are that the toxin is sequestered in either endosomes, Golgi-associated vesicles, or both. Unfortunately, there are no known natural endosomal markers. The Golgi apparatus in the LM cell is poorly developed and often difficult to discern. We postulate that at 19°C, the toxin is internalized by RME, trafficked via endosomes to the Golgi region, and routed to post-Golgi endosomes. Fusion with the lysosomal compartment is inhibited because 19°C is below the phase transition temperature for the lysosomal membrane. At 15°C the same events occur, but trafficking is inhibited prior to entry into the Golgi-associated vesicle because this temperature is lower than the phase transition for Golgi membranes. Taken together with the observation that it requires 3 logs more PE to kill cells at 15°C than at 19°C (TCD$_{50}$ is 9.0 and >10,000 ng, respectively), we postulate a role for the Golgi apparatus in the intoxication process.

c. Role of the Golgi Apparatus

We hypothesize that it is in the *trans*-Golgi region that the protoxin is nicked and activated. We envision that the mechanism by which this occurs is analogous to the proteolytic conversion of proproteins to proteins.[108-110] Most peptide hormones and neurotransmitters are synthesized as preproproteins that are proteolytically cleaved to yield active proteins. Examples include progastrin, proglucagon, procalcitonin, prorelaxin, promelanocortin, proenkephalin, provasopressin, prooxytocin, proparathyroid hormone, and proalbumin.[109] For example, the hormone insulin is synthesized on the rough endoplasmic reticulum as preproinsulin in the beta cells of islet of Langerhans of the pancreas. The pre-piece is removed in the lumen of the rough endoplasmic reticulum.[111] The proinsulin is subsequently routed to and through the Golgi apparatus, accumulates in secretory granules, and is released by a secretory process via exocytosis. During the intracellular progression from the endoplasmic reticulum to the cell surface, the proinsulin is modified to yield insulin. The processing involves the excision of a 9000-Da internal fragment of the proinsulin peptide and is mediated by a proprotein-cleaving enzyme. The excision process takes place either as a very late Golgi event (*trans*-most region) or in the secretory granule.[108,109] Recent data from Orci et al.[112,113] suggest that the proteolytic maturation of insulin is a post-Golgi event which occurs in acidic, clathrin-coated secretory vesicles.

A similar pattern of proprotein processing is highly conserved in nature with examples ranging from single cell eucaryotes to man.[114] A common feature characteristic of proprotein-processing enzymes is the recognition of paired basic amino acid sites, specifically Arg-Arg and Lys-Arg sites. The proprotein-cleaving enzyme (or family of enzymes) is not restricted to secretory cells as it has been well established that yeast[115,116] and liver cells[117,118] also process proproteins; i.e., pro-α-factor and proalbumin, respectively. These results suggest that all cells may have "proprotein-cleaving enzymes". Recently a calcium-dependent thiol protease isolated from yeast (KEX2) has been shown to meet the requirements for a bonafide "proprotein-cleaving enzyme" and has been used to process several mammalian proproteins.[114,119]

PE has two paired basic amino acid regions at position 182-183 (Arg-Arg) and 185-186 (Lys-Arg).[5] By analogy, DT also contains several paired basic amino acid regions: Lys-Arg sites at positions 125-126 and 172-173, and an Arg-Arg site at positions 192-193. The latter arginine-rich region connects the A and B domains and is readily cleaved *in vitro* by trypsin-like enzymes. A similar cleavage is thought to occur *in vivo* resulting in DT activation.[120] We speculate that the intracellular site of toxin activation both DT and PE is the *trans*-most region of the Golgi apparatus and is mediated by "proprotein-cleaving enzymes". It would be interesting to test the ability of KEX2 to nick PE in an *in vitro* assay.

To address the question of toxin activation on a biochemical level, we have used subcellular fractionation techniques. The protocol of these experiments is as follows. Harvest 2-d-old LM cell monolayers (never exposed to toxin), homogenize, and separate the fractions on the basis of buoyant density using ultracentrifugation in conjunction with Percoll density gradients. Isolated fractions are assayed for the ability to activate PE. Gradient fractions are incubated with an equal volume of activation buffer at pH 5.3 for 30 min at 37°C. PE is added and further incubated at 37°C for 60 min. Finally, to measure the ability of the gradient fractions to activate PE, the pH is raised to 8.2 by the addition of Tris buffer; wheat germ as a source of EF-2 and ^{14}C-NAD$^+$ is added, and the mixture is incubated at 25°C for 60 min. Identification of the fraction is based upon marker enzymes. β-Glucuronidase and β-galactosidase serve as markers for the lysosomal compartment, galactosyl transferase as the *trans*-Golgi marker, and peroxidase-labeled concanavalin A bound at 4°C as the plasma membrane marker. The results show that activation of PE occurs in fractions enriched for *trans*-Golgi and plasma membrane-associated enzymes, and in light lysosomes. Similar stimulation is also noted in endosomal enriched fractions. Heavy lysosomal fractions are unable to activate toxin. Activation is seen only at acidic pH 5.3. These data support the role of a nonlysosomal acidic organelle in the activation of PE.[121]

An alternative hypothesis, consistent with the data, is that toxin activation occurs in the endosomal compartment without the need for Golgi involvement. This would suggest that a low pH-induced conformational charge is sufficient to cause toxin activation or that the endosomal compartment contains proteases. Recent reports by several groups have described endosomal-associated proteases. Diment and Stahl[45] described an endosomal-associated cathepsin D-like protease which is responsible for degradation of ^{125}I-mannose-bovine serum albumin in rabbit alveolar macrophages. Schaudies et al.[46] show that EGF is proteolytically processed in A31 cells (a clone of mouse 3T3 fibroblasts) in the endosomal compartment. Further, the latter group describes several distinct endosomal subpopulations each of which contains different proteases.

Although we cannot at the present time exclude the possibility of toxin activation in the endosomal compartment, we favor the model implicating the Golgi apparatus as the site of activation for the following reasons. First, if as suggested by others, intracellular trafficking to and through the Golgi apparatus is inhibited at <15°C,[105,122] our temperature studies would suggest the Golgi as the site of activation. Second, although EF-2 isolated from mouse fibroblasts can be ADP-ribosylated by equal molar concentrations of PE and DT,[68,123] it requires 10^4 more DT molecules to kill mouse fibroblasts even though DT binds to mouse fibroblasts.[50,68] Following binding by LM cells, DT is routed to the lysosomal compartment via endosomes, and significant levels are seen in the endosomal compartment for up to 15 min following internalization.[71] Thus, if activation and escape were endosomal events, then DT should be toxic for mouse LM cells. Third, PE is routed in LM cells to the Golgi region shortly preceding the earliest expression of toxicity.[54,71] DT, on the other hand, is not routed to the Golgi region following internalization by LM cells.[70] Fourth, the Moehrings[50,124] have isolated a series of mutant Chinese Hamster ovary cells with differential sensitivity to PE, diphtheria toxin, and several viruses (see Chapter 3). One group of mutants, PV^r, is resistant to PE and to three enveloped RNA viruses but is sensitive to diphtheria toxin. These cells bind and internalize Sindbis virus normally; they also synthesize viral RNA as do wild-type cells. However, the mutant CHO cells do not process viral structural protein properly; specifically they cannot cleave PE2 to the mature E2 form. Presumably this cleavage occurs in the Golgi or in post-Golgi vesicles. It is possible that the same deficiency in processing renders these cells unable to activate PE, and thus resistant to the toxin. Our conclusion, based on these observations, is that activation of PE is a Golgi-related event. In keeping with the observation that activation occurs in an acidic compartment,[121] Anderson and Pathak[125] have reported that the vesicles and cisternae of the *trans* Golgi in human fibroblasts are acidic compartments.

In a series of recent reports, Olsnes and colleagues have presented evidence suggesting a role for the translocation of ricin into the cytosol via the *trans*-Golgi network (TGN).[126-128] Ricin is a plant toxin that shares structural similarities with PE.[129] van Deurs et al.[127] followed the internalization and routing of a monovalent ricin-horseradish peroxidase

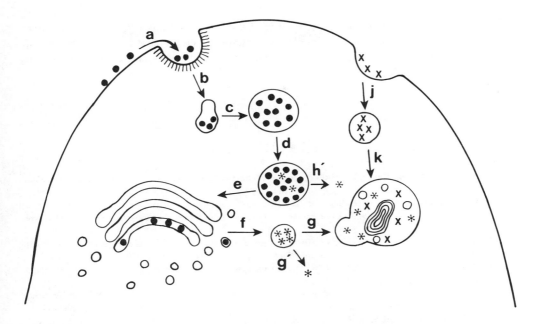

FIGURE 5.   Internalization and intracellular trafficking of PE (●) and DT (X) by mouse LM fibroblasts. (a) PE binds to cell surface receptors and enters by receptor-mediated endocytosis via clathrin-coated pits (30 s); (b, c, and d) PE is progressively routed through the endosomal compartment to the Golgi region (1 min to 10 min); (e and f) PE is delivered to the Golgi, activated (*), and exits from the *trans*-Golgi (10 to 20 min); (g) majority of PE is trafficked to the lysosomal compartment for degradation (≥ 20 min); (g') some of the activated PE escapes from a post-Golgi vesicle; (h') an alternative site of activation and escape is the endosomal compartment; (j) internalization of surface bound DT through non-clathrin-coated regions; (k) delivery of internalized DT directly to the lysosomal compartment without any Golgi involvement (≥ 5 min).

conjugate (Ri-HRP) in mammalian cells at 18 and 37°C. At the higher temperature, Ri-HRP is seen within the Golgi elements and toxicity is expressed. At 18°C, no Ri-HRP is seen in the Golgi and toxicity is inhibited. This has lead them to suggest that ricin may be modified by enzymes present in the Golgi.[127] Recently van Deurs et al.[128] have shown that ≃ 5% of the ricin internalized by mammalian cells reaches the Golgi after 60 min at 39.5°C, with about 70 to 80% of that being in the TGN. They suggest, on the basis of a combined morphological and biochemical approach, that translocation of the toxic ricin A chain to the cytosol occurs in the TGN. Similar results were obtained by Youle and Colombatti,[130] who studied the ricin intoxication of antiricin-producing hybridoma cells.

V. CONCLUDING REMARKS

In summary, the intoxication of sensitive mammalian cells by PE can be viewed as a multistep process involving (1) binding of the toxin to specific receptors on the cell surface, (2) internalization through clathrin-coated regions of the membrane, (3) intracellular trafficking and activation of the toxin, (4) escape of the enzymatically active polypeptide from an acidic, prelysosomal vesicle, and (5) ADP-ribosylation of cytoplasmic EF-2. The first, second, and fifth steps have been clearly defined. Evidence suggests escape from a Golgi-associated vesicle after activation by Golgi-associated enzymes. A diagram of our model showing the cellular basis for sensitivity of LM cells to PE is presented in Figure 5.

ACKNOWLEDGMENTS

The work presented in this manuscript was supported by funds provided by the National Institutes of Health (AI 17529 and GM 24028). The author would also like to express his appreciation to the *Journal of Histochemistry and Cytochemistry* and *Infection and Immunity* for permitting the use of previously published materials. The preparation of this manuscript would not have been possible without the thoughtful suggestions and editorial review by Dr. Catharine B. Saelinger.

REFERENCES

1. **Iglewski, B. H. and Kabat, D.,** NAD-dependent inhibition of protein synthesis by *Pseudomonas aeruginosa* toxin, *Proc. Natl. Acad. Sci. U.S.A.,* 72, 2284, 1975.
2. **Middlebrook, J. L. and Dorland, R. B.,** Response of cultured mammalian cells to the exotoxins of *Pseudomonas aeruginosa* and *Corynebacterium diphtheriae:* differential cytotoxicity, *Can. J. Microbiol.,* 23, 183, 1977.
3. **Leppla, S. H., Martin, O. C., and Muehl, L. A.,** The exotoxin of *P. aeruginosa*: A proenzyme having an unusual mode of activation, *Biochem. Biophys. Res. Commun.,* 81, 532, 1978.
4. **Middlebrook, J. L. and Droland, R. B.,** Bacterial toxins: Cellular mechanisms of action, *Microbiol. Rev.,* 49, 199, 1984.
5. **Gray, G. L., Smith, D. H., Baldridge, J. S., Harkins, R. N., Vasil, M. L., Chen, E. Y., and Heyneker, H. L.,** Cloning, nucleotide sequence, and expression in *Escherichia coli* of the exotoxin A structural gene of *Pseudomonas aeruginosa, Proc. Natl. Acad. Sci. U.S.A.,* 81, 2645, 1984.
6. **Allured, V. S., Collier, R. J., Carroll, S. F., and McKay, D. B.,** Structure of exotoxin A of *Pseudomonas aeruginosa* at 3.0-Angstrom resolution, *Proc. Natl. Acad. Sci. U.S.A.,* 83, 1320, 1986.
7. **Hwang, J., FitzGerald, D. J., Adhya, S., and Pastan, I.,** Functional domains of pseudomonas exotoxin identified by deletion analysis of the gene expressed in *E. coli, Cell,* 48, 129, 1987.
8. **Guidi-Rontani, C. and Collier, R. J.,** Exotoxin A of *Pseudomonas aeruginosa*: Evidence that domain 1 functions is receptor binding, *Mol. Microbiol.,* 1, 67, 1987.
9. **Jinno, Y., Chaudhary, V. K., Kondo, T., Adhya, S., FitzGerald, D. J., and Pastan, I.,** Mutational analysis of domain I of *Pseudomonas* exotoxin: Mutations in domain I of pseudomonas exotoxin which reduce cell binding and animal toxicity, *J. Biol. Chem.,* 263, 000, 1988.
10. **Saelinger, C. B., Snell, K., and Holder, I. A.,** Experimental studies on the pathogenesis of infections due to *Pseudomonas aeruginosa*: direct evidence for toxin production during pseudomonas infections of burned skin tissues, *J. Infect. Dis.,* 136, 555, 1977.
11. **Chaudhary, V. K., Xu, Y.-H., FitzGerald, D., Adhya, S., and Pastan I.,** Role of domain II of *Pseudomonas* exotoxin in the secretion of proteins into the periplasm and medium by *Escherichia coli, Proc. Natl. Acad. Sci. U.S.A.,* 85, 2939, 1988.
12. **Douglas, C. M. and Collier, R. J.,** Exotoxin A of *Pseudomonas aeruginosa*: Substitution of glutamic acid 553 with aspartic acid drastically reduces toxicity and enzymatic activity, *J. Bacteriol.,* 169, 4967, 1987.
13. **Carroll, S. F. and Collier, R. J.,** Active site of *Pseudomonas aeruginosa* exotoxin A. Glutamic acid 553 is photolabeled by NAD and show functional homology with glutamic acid 148 of diphtheria toxin, *J. Biol. Chem.,* 262, 8707, 1987.
14. **Carroll, S. F. and Collier, R. J.,** NAD binding site of diphtheria toxin: Identification of a residue within the nicotinamide subsite by photochemical modification with NAD, *Proc. Natl. Acad. Sci., U.S.A.,* 81, 3307, 1984.
15. **Carroll, S. F. and Collier, R. J.,** Amino acid sequence homology between the enzymic domains of diphtheria toxin and *Pseudomonas aeruginosa* exotoxin A, *Mol. Microbiol.,* 2, 293, 1988.
16. **Brown, M. S., Anderson, R. G. W., Basu, S. K., Goldstein, J. L.,** Recycling of cell-surface receptors: Observations from the LDL receptor system, *Cold Spring Harbor Symp. Quant. Biol.,* 46, 713, 1982.
17. **Goldstein, J. L., Brown, M. S., Anderson, R. G. W., Russell, D. W., Schneider, W. J.,** Receptor-mediated endocytosis: concepts emerging from the LDL receptor system, *Annu. Rev. Cell Biol.,* 1, 1, 1985.
18. **Pearse, B. M. F.,** Clathrin and coated vesicles, *EMBO J.* 6, 2507, 1987.
19. **Helenius, A., Mellman, I., Wall, D., and Hubbard, A.,** Endosomes, *Trends Biochem. Sci.,* 8, 245, 1983.
20. **Marsh, M., Bolzau, E., and Helenius, A.,** Penetration of Semliki Forest Virus from acidic prelysosomal vacuoles, *Cell,* 32, 931, 1983.
21. **Marsh, M., Griffiths, G., Dean, G. E., Mellman, I., and Helenius, A.,** Three dimensional structure of endosomes in BHK-21 cells, *Proc. Natl. Acad. Sci. U.S.A.,* 83, 2899, 1986.

22. **Paavola, L. G., Strauss, J. F., Boyd, C. O., and Nestler, J. E.,** Uptake of gold-and [³H] cholesteryl linoleate-labeled human low density lipoprotein by cultured rat granulosa cells: cellular mechanisms involved in lipoprotein metabolism and their importance to steroidogenesis, *J. Cell Biol.,* 100, 1235, 1985.
23. **Mellman, I., Fuchs, R., and Helenius, A.,** Acidification of the endocytic and exocytic pathways, *Annu. Rev. Biochem.,* 55, 663, 1986.
24. **Galloway, C. J., Dean, G. E., Marsh, M., Rudnick, G., and Mellman, I.,** Acidification of macrophage and fibroblast endocytic vesicles *in vitro, Proc. Natl. Acad. Sci. U.S.A.,* 80, 3334, 1983.
25. **Tyco, B. and Maxfield, F. R.,** Rapid acidification of endocytic vesicles containing α_2-macroglobulin, *Cell,* 28, 643, 1982.
26. **Goldstein, J. L., Anderson, R. G. W., and Brown, M. S.,** Coated pits, coated vesicles, and receptor mediated endocytosis, *Nature,* 279, 679, 1979.
27. **Geuze, H. J., Slot, J. W., Strous, G. J. A. M., Lodish, H. F., and Schwartz, A. L.,** Intracellular site of asialoglycoprotein receptor-ligand uncoupling: Double-label immunoelectron microscopy during receptor-mediated endocytosis, *Cell,* 32, 277, 1983.
28. **Geuze, H. J., Slot, J. W., Strous, G. J. A. M., Hasilik, A., and von Figura, K,** Ultrastructural localization of the mannose-6-phosphate receptor in rat liver, *J. Cell Biol.,* 98, 2047, 1984.
29. **Ciechanover, A., Schwartz, A. L., and Lodish, H. F.,** Sorting and recycling of cell surface receptors and endocytosed ligands: The asialoglycoprotein and transferrin receptors, *J. Cell Biochem.,* 23, 107, 1983.
30. **Dautry-Varsat, A., Ciechanover, A., and Lodish, H.,** pH and the recycling of transferrin during receptor-mediated endocytosis, *Proc. Natl. Acad. Sci. U.S.A.,* 80, 2258, 1983.
31. **Carpenter, G. and Cohen, S.,** ¹²⁵I-labeled human epidermal growth factor. Binding, internalization, and degradation in human fibroblasts, *J. Cell Biol.,* 71, 159, 1976.
32. **Carpenter, G., King, L., and Cohen, S.,** Epidermal growth factor stimulates phosphorylation in membrane preparations, *in vitro, Nature,* 276, 409, 1978.
33. **Kasuga, M., Karlsson, F. A., and Kahn, C. R.,** Insulin stimulates the phosphorylation of the 95,000-dalton subunit of its own receptor, *Science,* 215, 185, 1981.
34. **Carpentier, J. L., Gorden, P., Barazzone, P., Freychet, P., LeCam, A., and Orci, L.,** Intracellular localization of ¹²⁵I-labeled insulin in hepatocytes from intact rat liver, *Proc. Natl. Acad. Sci. U.S.A.,* 76, 2803, 1979.
35. **Mellman, J. and Plutner, H.,** Internalization and degradation of macrophage Fc receptors bound to polyvalent immune complexes, *J. Cell Biol.,* 98, 1170, 1984.
36. **Willingham, M. C., Haigler, H. T., FitzGerald, D. J. P., Gallo, M. G., Rutherford, A. V., and Pastan, I.,** The morphologic pathway of binding and internalization of epidermal growth factor in cultured cells. Studies on A431, KB, and 3T3 cells using multiple labelling methods, *Exp. Cell Res.,* 145, 163, 1983.
37. **Dunn, W. A. and Hubbard, A. L.,** Receptor-mediated endocytosis of epidermal growth factor by hepatocytes in the perfused rat liver: Ligand and receptor dynamics, *J. Cell Biol.,* 98, 2148, 1984.
38. **DiPaola, M. and Maxfield, F. R.,** Conformational changes in the receptor for epidermal growth and asialoglycoproteins induced by the mildly acidic pH found in endocytic vesicles, *J. Biol. Chem.,* 259, 9163, 1984.
39. **Burwen, S. J., Barker, M. E., Goldman, I. S., Hradek, G. T., Raper, S. E., and Jones, A. L.,** Transport of epidermal growth factor by rat liver: Evidence for a nonlysosomal pathway, *J. Cell Biol.,* 99, 1259, 1984.
40. **Beguinot, L., Lyall, R. M., Willingham, M. C., and Pastan, I.,** Down-regulation of the epidermal growth factor receptor in KB cell is due to receptor internalization and subsequent degradation in lysosomes, *Proc. Nat. Acad. Sci. U.S.A.,* 81, 2384, 1984.
41. **Renston, R. H., Jones, A. L., Christiansen, W. D., Hradek, G. T., and Underdown, B. J.,** Evidence for vesicular transport mechanism in hepatocytes for biliary secretion of immunoglobulin A, *Science,* 208, 1276, 1980.
42. **Geuze, H. J., Slot, J. W., Strous, G. J. A. M., Peppard, J., von Figura, K., Hasilik, A, and Schwartz, A. L.,** Intracellular receptor sorting during endocytosis: comparative immunoelectron microscopy of multiple receptors in rat liver, *Cell,* 37, 195, 1984.
43. **Kuhn, L. C. and Kraehenbuhl, J. P.,** Role of secretory component, a secreted glycoprotein, in the specific uptake of IgA dimer by epithelial cells, *J. Biol. Chem.,* 254, 11072, 1979.
44. **Mostov, K. E. and Blobel, G.,** A transmembrane precursor of secretory component, *J. Biol. Chem.,* 257, 11816, 1982.
45. **Diment, S. and Stahl, P.,** Macrophage endosomes contain proteases which degrade endocytosed protein ligands, *J. Biol. Chem.,* 260, 15311, 1985.
46. **Schaudies, R. P., Gorman, R. M., Savage, C. R., Jr., and Poretz, R. O.,** Proteolytic processing of epidermal growth factor within endosomes, *Biochem. Biophys. Res. Commun.,* 143, 710, 1987.
47. **Saelinger, C. B.,** ADP-ribosylating toxin, in *Bacterial Enzymes and Virulence,* Holder, I. A., Ed., CRC Press, Boca Raton, FL, 1985, 17.

48. **FitzGerald, D., Morris, R. E., and Saelinger, C. B.,** Receptor-mediated internalization of pseudomonas toxin by mouse fibroblasts, *Cell,* 21, 867, 1980.

49. **Manhart, M. D., Morris, R. E., Bonventre, P. F., Leppla, S., and Saelinger, C. B.,** Evidence for pseudomonas exotoxin A receptor on plasma membrane of toxin-sensitive LM fibroblasts, *Infect. Immun.,* 45, 596, 1984.

50. **Moehring, J. M. and Moehring, T. J.,** Strains of CHO-Kl cells resistant to Pseudomonas exotoxin A and cross-resistance to diphtheria toxin and viruses, *Infect. Immun.,* 41, 998, 1983.

51. **Tietze, C., Schlesinger, P., and Stahl, P.,** Chloroquine and ammonium ion inhibit receptor-mediated endocytosis of mannose-glycoconjugates by macrophages: apparent inhibition of receptor recycling, *Biochem. Biophys. Res. Commun.,* 93, 1, 1980.

52. **Schreiber, A. B., Schlessinger, J., Edidin, M.,** Interaction between major histocompatibility complex antigens and epidermal growth factor receptors on human cells, *J. Cell Biol.,* 98, 725, 1984.

53. **Morris, R. E. and Saelinger, C. B.,** Visualization of intracellular trafficking: Use of biotinylated ligands in conjunction with avidin-gold colloids, *J. Histochem. Cytochem.,* 32, 124, 1984.

54. **Morris, R. E. and Saelinger, C. B.,** Reduced temperature alters *Pseudomonas* exotoxin A entry into the mouse LM cell, *Infect. Immun.,* 52, 445, 1986.

55. **FitzGerald, D., Morris, R. E., and Saelinger, C. B.,** Essential role of calcium in cellular internalization of *Pseudomonas* toxin, *Infect. Immun.,* 35, 715, 1982.

56. **Morris, R. E.,** unpublished observations, 1986.

57. **Weber, B., Nickol, M., Jagger, K., and Saelinger, C. B.,** Interaction of pseudomonas exoproducts with phagocytic cells, *Can. J. Microbiol.,* 28, 679, 1982.

58. **Morris, R. E., Manhart, M. D., and Saelinger, C. B.,** Receptor mediated entry of *Pseudomonas* toxin: methylamine blocks clustering step, *Infect. Immun.,* 40, 806, 1983.

59. **Morris, R. E.,** Receptor-mediated endocytosis is required for expression of *Pseudomonas* and diphtheria toxin activity, in *Microbiology — 1985,* Levine, L., Ed., American Society for Microbiology, Washington, D.C., 1985, 91.

60. **Morris, R. E. and Saelinger, C. B.,** Route of *Pseudomonas* and diphtheria toxin entry into mammalian cells: basis for susceptibility to toxin, *Surv. Synth. Pathol. Res.,* 4, 34, 1985.

61. **Saelinger, C. B., Morris, R. E., and Foertsch, G.,** Trafficking of *Pseudomonas* exotoxin A in mammalian cells, *Eur. J. Clin. Microbiol.,* 4, 170, 1985.

62. **Saelinger, C. B. and Morris, R. E.,** Intracellular trafficking of *Pseudomonas* exotoxin A, *Antibiot. Chemother.,* 39, 149, 1987.

63. **Steinman, R. M., Silver, J. M., and Cohn, Z. A.,** Pinocytosis in fibroblasts. Quantitative studies, *in vitro, J. Cell Biol.,* 63, 949, 1974.

64. **Storrie, B., Pool, R. R., Sachdeva, M., Maurey, K. M., and Oliver, C.,** Evidence for both prelysosomal and lysosomal intermediates in endocytic pathways, *J. Cell Biol.,* 98, 108, 1984.

65. **van Deurs, B., Tønnessen, T. I., Peterson, O. W., Sandvig, K., and Olsnes, S.,** Routing of internalized ricin and ricin conjugates to the Golgi complex, *J. Cell Biol.,* 102, 37, 1986.

66. **Neurta, M. R., Ciechanover, A., Owen, L. S., and Lodish, H. F.,** Intracellular transport of transferrin- and asialoorosomucoid-colloidal gold conjugates to lysosomes after receptor-mediated endocytosis, *J. Histochem. Cytochem.,* 33, 1134, 1985.

67. **Willingham, M. C., Hanover, J. A., Dickson, R. B., and Pastan, I.,** Morphologic characterization of the pathway of transferrin endocytosis and recycling in human KB cells, *Proc. Nat. Acad. Sci. U.S.A.,* 81, 175, 1984.

68. **Heagy, W. E. and Neville, D. M.,** Kinetics of protein synthesis inactivation by diphtheria toxin in toxin-resistant L cells, *J. Biol. Chem.,* 256, 12788, 1981.

69. **Collins, D. and Huang, L.,** Cytotoxicity of diphtheria toxin A fragment to toxin-resistant murine cells delivered by pH-sensitive immunoliposomes, *Cancer, Res.,* 47, 735, 1987.

70. **Ohkuma, S. and Poole, B.,** Fluorescence probe measurement of the intralysosomal pH in living cells and the perturbation of pH by various agents, *Proc. Natl. Acad. Sci. U.S.A.,* 75, 3327, 1978.

71. **Morris, R. E. and Saelinger, C. B.,** Diphteria toxin does not enter resistant cells by receptor-mediated endocytosis, *Infect. Immun.,* 42, 812, 1983.

72. **Morris, R. E., Gerstein, A. S., Bonventre, P. F., and Saelinger, C. B.,** Receptor-mediated entry of diphtheria toxin into monkey kidney (Vero) cells: Electron microscopic evaluation, *Infect. Immun.,* 50, 721, 1985.

73. **FitzGerald, D., Morris, R. E., and Saelinger, C. B.,** Inhibition of *Pseudomonas* toxin internalized by methylamine, *Rev. Infect. Dis.,* 5, S985, 1983.

74. **Marshall, S. and Olefsky, J. M.,** Effects of lysosomotropic agents on insulin interactions with adipocytes. Evidence for a lysosomal pathway for insulin processing and degradation, *J. Biol. Chem.,* 254, 10153, 1979.

75. **van Leuven, F., Cassiman, J.-J., and Van Den Berghe, H.,** Primary amines inhibit recycling of α_2M receptors in fibroblasts, *Cell,* 20, 37, 1980.

76. **King, A. C., Hernaez-Davis, L., Cuatrecasas, P.,** Lysosomotropic amines cause intracellular accumulation of receptors for epidermal growth factor, *Proc. Natl. Acad. Sci. U.S.A.,* 77, 3283, 1980.
77. **Schwartz, A. L., Bolognesi, A., and Fridovich, S. E.,** Recycling of the asialoglycoprotein receptor and the effect of lysosomotropic amines in hepatoma cells, *J. Cell Biol.,* 98, 732, 1984.
78. **Sundan, A., Sandvig, K., and Olsnes, S.,** Calmodulin antagonists sensitize cells to pseudomonas toxin, *J. Cell. Physiol.,* 119, 15, 1984.
79. **Sundan, A., Sandvig, K., and Olsnes, S.,** Effect of malignant transformation, retinoic acid, trifluoperazine and W7 on the sensitivity of cells to pseudomonas toxin, *Cancer Res.,* 44, 4919, 1984.
80. **Dumont, M. E. and Richards, F. M.,** The pH dependent conformational change of diphtheria toxin, *J. Biol. Chem.,* 263, 2087, 1988.
81. **Blewitt, M. G., Chung, L. A., and London, E.,** Effect of pH on the conformation of diphtheria toxin and its implications for membrane penetration, *Biochemistry,* 24, 5458, 1985.
82. **Sandvig, K. and Olsnes, S.,** Rapid entry of nicked diphtheria toxin into cells at low pH. Characterization of the entry process and effects of low pH on the toxin molecule, *J. Biol. Chem.,* 256, 9068, 1981.
83. **Sandvig, K. and Olsnes, S.,** Diphtheria toxin entry into cells is facilitated by low pH, *J. Cell. Biol.,* 87, 828, 1980.
84. **Draper, R. K. and Simon, M. I.,** The entry of diphtheria toxin into the mammalian cell cytoplasm: evidence for lysosomal involvement, *J. Cell Biol.,* 87, 849, 1980.
85. **Kagan, B. L., Finkelstein, A., and Colombini, M.,** Diphtheria toxin fragment forms large pores in phospholipid bilayer mambranes, *Proc. Natl. Acad. Sci. U.S.A.,* 78, 4950, 1981.
86. **Donovan, J. J., Simon, M. I. and Montal, M.,** Insertion of diphtheria toxin into and across membranes: role of phosphoinositide asymmetry, *Nature,* 298, 669, 1982.
87. **Donovan, J. J., Simon, M. I., Draper, R. K., and Montal, M.,** Diphtheria toxin forms transmembrane channels in planar lipid bilayers, *Proc. Natl. Acad. Sci. U.S.A.,* 78, 172, 1981.
88. **Donovan, J. J., Simon, M. I., and Montal, M.,** Requirements for the translocation of diphtheria toxin fragment A across lipid membranes, *J. Biol. Chem.,* 260, 8817, 1985.
89. **Hoch, D. H., Romero-Mira, M., Ehrlich, B. E., Finkelstein, A., DasGupta, B. R., and Simpson, L. L.,** Channels formed by botulinum, tetanus, and diphtheria toxins in planar lipid bilayers: relevance to translocation of proteins across membranes, *Proc. Nat. Acad. Sci. U.S.A.,* 82, 1692, 1985.
90. **Donovan, J. J. and Middlebrook, J. L.,** Ion-conducting channels produced by botulinum toxin in planar lipid membranes, *Biochemistry,* 25, 2872, 1986.
91. **Simpson, L. L.,** Ammonium chloride and methylamine hydrochloride antagonize clostridial neurotoxins, *J. Pharmacol. Exp. Ther.,* 225, 546, 1983.
92. **Madshus, I. H., Olsnes, S., and Sandvig, K.,** Mechanism of entry into the cytosol of poliovirus type 1: requirement for low pH, *J. Cell Biol.,* 98, 1194, 1984.
93. **Madshus, I. H., Olsnes, S., and Sandvig, K.,** Requirements for entry of poliovirus RNA into cells at low pH, *EMBO J.,* 3, 1945, 1984.
94. **Zeichhardt, H., Wetz, K., Willingmann, P., and Habermehl, K.-O.,** Entry of poliovirus type 1 and Mouse Elberfeld (MD) Virus into HEp-2 cells: receptor-mediated endocytosis and endosomal or lysosomal uncoated, *J. Gen. Virol.,* 66, 483, 1985.
95. **FitzGerald, D. J. P., Trowbridge, I. S., Pastan, I., and Willingham, M. C.,** Enhancement of toxicity of antitransferrin receptor antibody *Pseudomonas* exotoxin conjugates by adenovirus, *Proc. Nat. Acad. Sci. U.S.A.,* 80, 4134, 1983.
96. **FitzGerald, D. J. P., Padmanabhan, P., Pastan, I., and Willingham, M. C.,** Adenovirus-induced release of epidermal growth factor and *Pseudomonas* toxin into the cytosol of KB cells during receptor-mediated endocytosis, *Cell,* 32, 607, 1983.
97. **Seth, P., FitzGerald, D. J. P., Willingham, M. C., and Pastan, I.,** Role of a low pH environment in adenovirus enhancement of the toxicity of a *Pseudomonas* exotoxin-epidermal growth factor conjugate, *J. Virol.,* 51, 650, 1984.
98. **Seth, P., FitzGerald, D., Willingham, M., and Pastan, I.,** Pathway of adenovirus entry into cells, in *Virus Attachment and Entry into Cells,* Crowell, R. L. and Lonberg-Holm, K., Eds., American Society for Microbiology, Washington, D.C., 1986, 191.
99. **Zalman, L. S. and Wisnieski, B. J.,** Characterization of the insertion of *Pseudomonas* exotoxin A into membranes, *Infect. Immun.,* 50, 630, 1985.
100. **Farahbakhsh, Z. T., Baldwin, R. L., and Wisnieski, B. J.,** *Pseudomonas* exotoxin A: membrane binding, insertion, and traversal, *J. Biol. Chem.,* 261, 11404, 1986.
101. **Farahbakhsh, Z. T., Baldwin, R. L., and Wisnieski, B. J.,** Effect of low pH on the conformation of *Pseudomonas* exotoxin A, *J. Biol. Chem.,* 262, 2256, 1987.
102. **Olsnes, S. and Sandvig, K.,** How proteins enter and kill cells, in *Immunotoxins,* Frankel, A. E., Ed., Martinus Nijhoff, Boston, 1988.
103. **Weigel, P. H. and Oka, J. A.,** The surface content of asialoglycoprotein receptors on isolated hepatocytes is reversibly modulated by changes in temperature, *J. Biol. Chem.,* 258, 5089, 1983.

104. **Dunn, W. A., Hubbard, A. L., and Aronson, N. N., Jr.,** Low temperature selectively inhibits fusion between pinocytic vesicles and lysosomes during heterophagy of ^{125}I-asialofetuin by the perfused rat liver, *J. Biol. Chem.,* 255, 5971, 1980.

105. **Sleight, R. G. and Pagano, R. E.,** Transport of a fluorescent phosphatidylcholine analog from the plasma membrane to the Golgi apparatus, *J. Cell Biol.,* 99, 742, 1984.

106. **Marnell, M. H., Shia, S.-P., Stookey, M., and Draper, R. K.,** Evidence for penetration of diphtheria toxin to the cytosol through a prelysosomal membrane, *Infect. Immun.,* 44, 145, 1984.

107. **Hudson, T. H. and Neville, D. M.,** Temporal separation of protein toxin translocation from processing events, *J. Biol. Chem.,* 262, 16484, 1987.

108. **Docherty, K., Steiner, D. F.,** Post-translational proteolysis in polypeptide hormone biosynthesis, *Annu. Rev. Physiol.,* 44, 625, 1982.

109. **Steiner, D. F., Docherty, K., and Carroll, R.,** Golgi/Granule processing of peptide hormone and neuropeptide precursors: a minireview, *J. Cell Biochem.,* 24, 121, 1984.

110. **Loh, Y. P., Brownstein, M. J., and Gainer, H.,** Proteolysis in neuropeptide processing and other neural functions, *Annu. Rev. Neurosci.,* 7, 189, 1984.

111. **Panzeh, C., Labrecque, A. D., Duguid, J. R., Carroll, R. J., Keim, P. S., Heinrikson, R. L., and Steiner, D. F.,** Detection and kinetic behavior of preproinsulin in pancreatic islets, *Pro. Natl. Acad. Sci. U.S.A.,* 75, 1260, 1978.

112. **Orci, L., Ravazzola, M., Storch, M.-J., Anderson, R. G. W., Vassalli, J.-D., and Perrelet, A.,** Proteolytic maturation of insulin is a post-Golgi event which occurs in acidifying clathrin-coated secretory vesicles, *Cell,* 49, 865, 1987.

113. **Orci, L., Halban, P., Amherdt, M., Ravazzola, M., Vassalli, J.-D., and Perrelet, A.,** A clathrin-coated, Golgi-related compartment of the insulin secreting cell accumulates proinsulin in the presence of monensin, *Cell,* 39, 39, 1984.

114. **Marx, J. L.,** A new wave of enzymes for cleaving prohormones, *Science,* 235, 285, 1987.

115. **Julius, D., Brake, A., Blair, L., Kunisawa, R., and Thorner, J.,** Isolation of the putative structural gene for the lysine-arginine-cleaving endopeptidase required for processing of yeast prepro-α-factor, *Cell,* 37, 1075, 1984.

116. **Julius, D., Schekman, R., and Thorner, J.,** Glycosylation and processing of prepro-α-factor through the yeast secretory pathway, *Cell,* 36, 309, 1984.

117. **Vlasuk, G. P., Ghrayeb, J., and Walz, F. G., Jr.,** Proalbumin is bound to the membrane of rat liver smooth microsomes, *Biochem. Biophys. Res. Commun.,* 94, 366, 1980.

118. **Judah, J. D. and Nicholls, M. R.,** Biosynthesis of rat serum albumin, *Biochem. J.,* 123, 649, 1971.

119. **Bathurst, I. C., Brennan, S. O., Carrell, R. W., Cousens, L. S., Brake, A. J., and Barr, P. J.,** Yeast KEX2 protease has the properties of a human proalbumin converting enzyme, *Science,* 235, 348, 1987.

120. **Greenfield, L., Bjorn, M. J., Horn, G., Fong, D., Buck, G. A., Collier, R. J., and Kaplan, D. A.,** Nucleotide sequence of the structural gene for diphtheria toxin carried by corynebacteriophage β, *Proc. Nat. Acad. Sci. U.S.A.,* 80, 6853, 1983.

121. **Kozak, K., Morris, R. E., and Saelinger, C. B.,** in revision.

122. **Saraste, J., Kuismanen, E.,** Pre- and post-Golgi vacuoles operate in the transport of Semliki Forest Virus membrane glycoproteins to the cell surface, *Cell,* 38, 535, 1984.

123. **Moehring, T. J. and Moehring, J. M.,** Selection and characterization of cells resistant to diphtheria toxin and *Pseudomonas* exotoxin A: presumptive translational mutants, *Cell,* 11, 447, 1977.

124. **Didsbury, J. K., Moehring, J. M., and Moehring, T. J.,** Binding and uptake of diphtheria toxin by toxin-resistant Chinese Hamster ovary and mouse cells, *Mol. Cell Biol.,* 3, 1283, 1983.

125. **Anderson, R. G. W. and Pathak, R. K.,** Vesicles and cisternae in the trans Golgi apparatus of human fibroblasts are acidic compartments, *Cell,* 40, 635, 1985.

126. **Sandvig, K., Tønnessen, T. I., Olsnes, S.,** Ability of inhibitors of glycosylation and protein synthesis to sensitize cells to abrin, ricin, *Shigella* toxin, and *Pseudomonas* toxin, *Cancer Res.,* 46, 6418, 1986.

127. **van Deurs, B., Sandvig, K., Petersen, O. W., Olsnes, S., Simon, K., and Griffiths, G.,** Estimation of the amount of internalized ricin that reaches the trans-Golgi network, *J. Cell Biol.,* 106, 253, 1988.

128. **van Deurs, B., Petersen, O. W., Olsnes, S., and Sandvig, K.,** Delivery of internalized ricin from endosomes to cisternal Golgi elements is a discontinuous, temperature-sensititve process, *Exp. Cell Res.,* 171, 137, 1987.

129. **Olsnes, S. and Phil, A.,** Chimeric toxins, *Pharmac. Ther.* 15, 335, 1982.

130. **Youle, R. J. and Colombatti, M.,** Hybridoma cells containing intracellular anti-ricin antibodies show ricin meets secretory antibodies before entering the cytosol, *J. Biol. Chem.,* 262, 4676, 1987.

131. **Morris, R. E. and Saelinger, C. B.,** in preparation.

Chapter 5

EXTRACELLULAR AND INTRACELLULAR TRAFFICKING OF CLOSTRIDIAL TOXINS

Lance L. Simpson

TABLE OF CONTENTS

I. INTRODUCTION

There are a host of protein toxins that are exquisitely potent and that share a number of structural and functional properties. Most of these toxins are microbial in origin, although several are derived from plants, and among their shared properties are the following.[1] They are large molecules, ranging in molecular mass from 50,000 to 150,000 daltons. Each molecule is composed of two or more functional domains, and thus the holotoxin is needed to produce effects on intact tissues. These protein toxins show a marked preference for eukaryotic cells as opposed to prokaryotic cells, they typically act at the internal face of the plasma membrane or in the cytosol, and they express their toxicity by virtue of being enzymes.

There are a number of toxins that are known to belong to this group, several of which have been discussed in other chapters in this book. In addition to these known representatives, there are additional toxins that may belong to the group, but evidence to support this claim is incomplete. Clostridial toxins are good examples of agents that are being intensively studied and that may belong to the broader group of potent toxins that act inside eukaryotic cells.

There are three classes of clostridial toxins that are of interest here: (1) the botulinum neurotoxins, (2) the clostridial binary toxins, and (3) tetanus toxin. The botulinum neuro-toxins, which act mainly at peripheral cholinergic nerve endings to block the release of acetylcholine, are generally considered the most poisonous substances known. Tetanus toxin acts mainly in the central nervous system to block the release of inhibitory transmitters, and it is viewed as the second most poisonous substance. The clostridial binary toxins are substances that have only recently been characterized, and the full spectrum of their target cell specificity and potency has not been determined. However, two points are worth noting. The binary toxins are not selective in acting on nerve cells, nor do they show cholinergic blocking properties. To the contrary, they appear to act on a variety of eukaryotic cell types, and the most dramatic effect of poisoning is cell lysis. Second, the binary toxins are the only substances considered here that have been shown conclusively to act inside cells. Each of the binary toxins has an enzymatic chain that possesses ADP-ribosyltransferase activity, and the intracellular substrate appears to be actin.

This chapter reviews the major features of the clostridial toxins, including their origin, structure, and presumed mechanism of action. The data that are available for the clostridial toxins are compared to that which is available for better-characterized toxins, such as diphtheria toxin and ricin. From these data and comparisons, tentative conclusions are drawn about the internalization and intracellular trafficking of clostridial toxins.

II. BOTULINUM NEUROTOXIN

A. MICROBIAL ORIGIN

Botulinum neurotoxin is produced in at least seven partially or wholly distinct serotypes, designated A,B,C,D,E,F, and G.[2] The neurotoxin is produced by the microorganism *Clostridium botulinum*, and there are several important points that relate to microbial growth and toxin production. To begin with, most strains of the bacteria produce only one serotype, but there are exceptions. In terms of increasing complexity, the exceptions can be enumerated as follows: (1) organisms that produce mainly one serotype, with only trace amounts of a second serotype (i.e., a strain that synthesizes predominantly type F and nominal amounts of type A;[3]) (2) organisms that make substantial amounts of two botulinum neurotoxins (i.e., numerous strains produce both type C and type D;[4]) and (3) organisms that synthesize both types C and D neurotoxin, and in addition produce a binary toxin (see below).

Another interesting point that bears on microbial growth and toxin production relates to the genetic material that encodes the toxin. To date, no investigator has localized an encoding sequence within the genome of the host organism. To the contrary, in the two cases in which the genetic material has been localized (types C and D) that material was viral in origin.[5-7] In one other case (type G), there is preliminary evidence that the genetic material may be plasmid mediated.[8]

The final point relates to microbial physiology. It appears that toxin production plays no essential role in the growth and reproduction of clostridia. For those organisms in which toxigenicity is linked to lysogenicity, curing the organisms halts the production of neurotoxin, but it has no obvious effect on microbial physiology. In related experiments, investigators have shown that microbial growth and toxin production can be manipulated separately and independently. These data indicate that although clostridia are responsible for toxin production, they do not rely on the toxin for any aspect of their metabolism.

There is additional evidence that the genetic material for serotypes other than C, D, and G may be under the control of viral, plasmid, or related extrachromosomal elements. In the recent past, outbreaks of botulism — the disease that is caused by botulinum neurotoxin — have been attributed to nonbotulinum organisms. Both *Clostridium berati* and *C. butyricum* have been shown to produce the neurotoxin.[9,10] One convenient way to account for this is that a virus or related particle that carries the genetic information for the toxin can infect or be transmitted to bacteria other than *C. botulinum*.

B. STRUCTURE OF THE TOXIN MOLECULE

There is microheterogeneity among the seven serotypes of botulinum neurotoxin, but for the moment the discussion will focus on patterns of similarity. The neurotoxin is synthesized as a single chain polypeptide with a molecule mass of approximately 150,000 (values range from 142,000 to 167,000). In the immediate posttranslational stage, the molecule has relatively little toxicity. When the toxin is exposed to microbial trypsin-like enzymes or to mammalian trypsin, it is cleaved to yield a dichain molecule in which a light chain ($M_r \sim 50,000$) is linked by a disulfide bond and by noncovalent forces to a heavy chain ($M_r \sim 100,000^2$). This is the active form of the molecule. Reduction of the disulfide bond renders the molecule nontoxic.[11] As this finding would suggest, the isolated light chain and the isolated heavy chain are individually nontoxic.

The nicked botulinum neurotoxin molecule can be further enzymatically cleaved to give two stable fragments. Mild digestion with papain produces a fragment ($M_r \sim 100,000$) that is composed of the light chain covalently bound to the amino-terminus of the heavy chain. The second fragment ($M_r \sim 50,000$) represents the carboxy-terminus of the heavy chain. A similar pattern of trypsin-induced nicking and subsequent papain-induced digestion is obtained with tetanus toxin (see below). Although the patterns of protein chemistry are the same, the nomenclature used by botulinum toxin workers and tetanus toxin workers is highly disparate. For the purposes of this review, the following terms will be employed. The 50,000-Da light chain and the 100,000-Da heavy chain will be referred to as such. The amino-terminus of the heavy chain will be labeled H_1 and the carboxy-terminus will be called H_2.

C. SYSTEMIC ACTIONS

There are three recognized etiologies for botulism, and these are referred to as food-borne botulism, infant botulism, and wound botulism. Food-borne botulism is a form of primary intoxication in which the patient ingests preformed toxin. Infant botulism and wound botulism are primary infections that lead secondarily to intoxication. The organism colonizes the gut or a wound, and toxin is synthesized and released within the patient.

In the case of wound botulism, extracellular and intracellular trafficking are of minor importance. The toxin probably diffuses from the wound to local endplates, or to the general circulation and from there to endplates throughout the body. In the case of food-borne botulism and infant botulism, the situation is more complex. The toxin must leave the gastrointestinal tract to reach lymphatics and blood, and this almost certainly requires that it penetrate membranes. Ironically, this limiting step in the onset of botulism has never been the subject of sustained study. No one has determined the cells that are crossed, the mechanism for binding (assuming there are receptors) and internalization, the route for movement of toxin from one cell surface to another, or the mechanism for releasing the toxin into the circulation. This is a potentially fertile area for the study of intracellular trafficking of botulinum neurotoxin, and the results would contribute greatly to an understanding of toxin action.

The clinical presentation of botulism is marked by neurological dysfunction. Among the more common features of the disease are (1) loss of pupillary reflexes, (2) dysarthria and dysphagia, (3) descending motor weakness, initially involving the cranial nerves and later involving the nerves of the trunk, and upper and lower extremities, (4) bilateral paralysis, including loss of function of the muscles of respiration (intercostal muscles and diaphragm), and (5) autonomic dysfunction, such as urinary retention and adynamic ileus. This array of problems led early workers to assume that the toxin produced a central nervous system disorder that had numerous peripheral manifestations,[12] but it is now known that the toxin acts exclusively in the peripheral nervous system.[13,14] All of the signs and symptoms listed above can be accounted for on the basis of a single mode of action. The toxin blocks cholinergic transmission, mainly at the neuromuscular junction but also at autonomic ganglia and at postganglionic parasympathetic sites.

There is implicit within these comments a point to be made about trafficking. Botulinum neurotoxin has no mechanism for being transported productively into the central nervous system. There is laboratory evidence for abortive retrograde transport of the toxin by motoneurons,[15,16] but this is not associated with blockade of synaptic transmission. Furthermore, the toxin is not known to penetrate the blood-brain barrier. Hence, its movement is confined to the periphery, and this is mediated by blood and lymph.

D. CELLULAR ACTIONS

All serotypes of botulinum neurotoxin block transmission at cholinergic neuroeffector sites. They show greatest potency when assayed on alpha-motoneurons, such as those that innervate the diaphragm, and thus isolated neuromuscular preparations have been the tissues of choice for study of these agents. Several decades of research on the toxin have culminated in the finding that the cellular mechanism of action is blockade of acetylcholine release. The toxin does not have, as its primary mechanism of action, blockade of the high affinity uptake of choline, inhibition of acetylcholine synthesis, disruption of vesicle filling with acetylcholine, or any postsynaptic effect, such as occlusion of the receptor or inhibition of acetylcholinesterase.[13,14]

One of the major impediments to developing a thorough understanding of the mechanism of toxin action has been the relatively poor base of knowledge that bears on excitation-secretion coupling. As the author has pointed out numerous times, the study of botulinum neurotoxin requires that one decipher pathophysiology at a level more molecular than our knowledge of normal physiology will support. This is indeed a challenge. However, there is at least one substance that has been consistently implicated in vesicular and granular exocytosis, this being calcium, and as a result a diproportionate amount of the literature deals with the interaction between toxin and calcium. This is explored more fully in a later section, but two points should be made here. In every tissue in which neurotransmitter or neurohormone secretion has been studied, calcium has been found to facilitate, or to be an absolute requirement for, the process.[17] However, those few studies that have examined the interaction between toxin and calcium agree that botulinum neurotoxin does not act at the plasma membrane to block depolarization-induced influx of calcium. There is a consensus that the toxin acts to block exocytosis, but at a step that is distal to the entry of calcium into secreting cells.

Most research now centers on the idea that the toxin blocks exocytosis by virtue of proceeding through a three-step sequence of events.[14,18,19] There is an initial binding step that produces no observable effects on transmitter release, a subsequent translocation step that is thought to involve internalization of the molecule, and a final poisoning step that is responsible for inhibiting exocytosis. Furthermore, there is suggestive evidence that the structure of the toxin molecule can be linked to this sequence of events.[19] The carboxy-terminus of the heavy chain appears to play an important role in the binding step, the amino-terminus of the heavy chain has been implicated in the internalization process, and the light chain is assumed to be responsible for intracellular poisoning.

E. INTRACELLULAR TRAFFICKING

Of the three steps thought to be involved in toxin action, the one of major concern here is internalization and subsequent trafficking. Indeed, one of the most pressing issues in clostridial toxin research is that of verifying the existence of an internalization step.

For many of the potent protein toxins, such as diphtheria toxin, cholera toxin, and the plant lectins, there is no question about internalization. The substrates for these toxins are found on the internal surface of the plasma membrane or in the cytoplasm. This virtually assures that the toxins, or some component of these toxins, enters the cell. In addition, there have been cleverly designed experiments that add to the evidence that these toxins exert their effects in the cell interior. For example, cell-fusion experiments have been used to show that the catalytic chains of certain toxins, if they are artificially introduced into the cell interior, will cause poisoning.[20,21] As another example, cell fusion has been used to introduce antibody into vulnerable cells, and this renders cells resistant to toxin added to the surrounding medium.[22] The inescapable conclusion is that toxin is imported into the cell, where it is antagonized by neutralizing antibody.

The body of compelling evidence that exists for agents like diphtheria toxin, ricin, and related substances does not exist for botulinum neurotoxin. It must be acknowledged straight-away that an intracellular substrate has not been identified, and only one experiment has been reported that truly implicates an intracellular site of action.[23] The bulk of the evidence is indirect. This may seem like a bleak state of affairs, but there are two matters that help to place the issue in a more favorable light. First, although there is no single piece of datum that proves directly that botulinum neurotoxin acts inside nerve cells, there is a considerable amount of indirect and suggestive evidence. The latter encourages many investigators to believe that an intracellular locus of action will eventually be identified. Secondly, botulinum neurotoxin seems to possess a notable similarity to diphtheria toxin in terms of structure-function relationships. Diphtheria toxin is known to act in the cell interior, where it ADP-ribosylates elongation factor 2.[24] By analogy, one might tentatively deduce that botulinum toxin also acts intracellularly to enzymatically inactivate a vital process.

The most instructive approach to evaluating the possibility that the toxin is internalized is to examine individually the several observations and arguments that support the hypothesis. These can be broken into three categories: (1) indirect evidence that is closely related to the process of neuromuscular blockade, (2) direct evidence that may be closely related to the onset of neuromuscular blockade, and (3) a single experiment on intracellular injection.

The first observation to support the hypothesis of internalization was indirect in nature.[18] The study involved the analysis of botulinum neurotoxin type A action on the isolated phrenic nerve-hemidiaphragm preparation. The work was an outgrowth of an earlier observation made by Burgen et al.,[25] who had noted that the process of binding did not appear to be synonymous with the process of neuroparalysis. They found that tissues could be exposed briefly to the toxin and then washed, but there was not an immediate onset of blockade. Instead, the onset of paralysis developed gradually, and it was proportional both to the length of exposure and to the concentration of toxin. Numerous authors have reproduced the finding, and they have interpreted it to mean that paralysis involves at least two steps, initial binding and later paralysis.

The idea that there is a third step, and that it involves internalization, stems from experiments with neutralizing antibody.[18] Tissues were incubated with toxin under conditions that would allow binding to proceed to completion but would retard any supposed internalization step (i.e., low temperature, low calcium, and no nerve stimulation). At the end of incubation, tissues were washed free of unbound toxin. They were then exposed to neutralizing antibody at various times after physiological conditions were restored (i.e., 35°C, 1.8 mM Ca^{2+}, with nerve stimulation). The data showed that the immediate application of neutralizing antibody substantially antagonized the onset of toxin-induced neuromuscular blockage. However, if the application of antibody was delayed for a short time, it no longer afforded any protection. The interpretation assigned to these findings was that the toxin attached with high affinity to receptors on the cell surface, but this binding left the molecule accessible to the neutralizing effects of antibody. When exposed to conditions that favored internalization, the toxin rapidly disappeared from accessibility to antibody. Furthermore, the apparent movement of the toxin molecule occurred long before onset of neuromuscular blockade. This could easily be explained by assuming that a membrane penetration step was interposed between the binding step and the poisoning step.

This observation was the forerunner to numerous apparently supportive findings and deductions, but before these are presented there should be a consideration of possible alternate explanations. The fact that the toxin disappears from accessibility to neutralizing antibody has been taken to mean that the toxin penetrates the membrane. Another possibility is that the toxin remains in place at the surface of the plasma membrane, but it undergoes a conformational change. Due to steric and/or other forms of hindrance, polyclonal antibody can no longer gain access to its epitopes. Alternatively, the toxin could insert into the membrane, and by virtue of this, the epitopes in the molecule would become shielded.

The possibility that the toxin penetrates the membrane and the possibility that it remains at the cell surface are the two that are easiest to discriminate experimentally and logically. Therefore, a series of observations and logical deductions are invoked to show that penetration is the more plausible explanation.

1. The suggestion that the toxin remains at the cell surface but undergoes a conformational change is not a convincing explanation for the finding with antibody. Clostridial neurotoxins are relatively large molecules ($M_r \sim 150,000$) that have numerous antigenic domains.[26] These antigenic sites occur both within chains and across chains. One might propose that a toxin molecule could undergo a conformational change that would occlude one or a small number of epitopes, but it is a dubious proposition that a large protein could undergo a conformational change that would abolish all antigenic domains. While not impossible, it is rather unlikely.

2. There is a lag time that intervenes between apparent disappearance of the toxin and onset of neuromuscular blockade. This is hard to explain if the toxin is acting at the cell surface, and especially hard to explain if that surface action is stoichiometric. Conversely, a lag time is easy to reconcile with the idea of an internalization process that is followed by an intracellular enzymatic action.

3. The rate of onset of neuromuscular blockade is nerve activity dependent.[27] When the rate of nerve stimulation is increased, the rate of onset of paralysis is also increased. There is abundant evidence that exocytosis and endocytosis are linked. This could mean that rapid rates of nerve stimulation promote onset of paralysis by promoting receptor-mediated endocytosis.

4. Botulinum neurotoxin is an exquisitely potent substance that acts rather selectively to block the release of acetylcholine. For the toxin to act so selectively and in such small

quantities, one or the other of two expectations should be met (1) the toxin acts stoichiometrically, and, therefore, of necessity would have to bind at sites that govern exocytosis, or (2) the toxin acts catalytically, and although this may not occur exactly at the site of exocytosis, the effects would be expressed in a way that alters exocytosis. As noted above and elsewhere,[14] the mechanism is most probably enzymatic. But this poses serious problems for the notion that the toxin acts at the cell surface. Black and colleagues have conducted ultrastructural studies on toxin binding to nerve endings.[16,28,29] They found that the toxin did bind preferentially to nerve endings, but it did not localize exclusively in the zones of exocytosis. This is compatible with the notion of endocytosis; the internalized molecule could exert an enzymatic effect on one step in a cascade of events that govern transmitter release. On the other hand, this is less compatible with a model that calls for the toxin to act at the cell surface. Botulinum neurotoxin would have to exert an enzymatic effect at sites remote from release zones, and the effect would have to "ripple" down the membrane to impact on the release zones. The credibility of such a scheme must be judged against the findings that the toxin has already been shown not to have direct effects on a variety of membrane functions, including the high affinity uptake of choline, the resting membrane potential, the flux of calcium across membranes, and the activity of Na^+-K^+ ATPase. Indeed, even release of transmitter across the membrane of poisoned cells is still possible, given the appropriate secretogogue.

It is difficult to envision how the toxin could act on the outside of the cell to produce an effect that would ripple down the membrane and yet leave virtually all aspects of membrane structure and function intact. Furthermore, the observation that certain secretogogues can evoke transmitter release from poisoned nerves would appear to necessitate the surprising conclusion that normal transmitter release and secretogogue-induced transmitter release involve different parts of the membrane. None of these problems arise within the framework of internalization. The fact that the toxin does not bind exclusively at release zones would be largely irrelevant. The salient issue is whether the toxin binds at sites that would accommodate the internalization process. Furthermore, the fact that certain secretogogues can trigger acetylcholine release from poisoned nerves would not be hard to explain. In all likelihood, there is a cascade of events that intervenes between depolarization of the nerve ending and eventual release of transmitter. It is conceivable that the toxin acts at some point in this intracellular cascade, and it is conceivable that certain secretogogues can bypass or act distally to the point of toxin attack.

5. In trying to determine the mechanism of action of clostridial toxins, it is useful to identify substances that can serve as models. Of the various substances whose mechanisms of action are known, none appears to be a better model than diphtheria toxin (see Chapter 3). The clostridial neurotoxins, both botulinum and tetanus, share with diphtheria toxin the following properties: (1) the toxins are microbial in origin: (2) the genetic material that encodes the toxin is not found in the bacterial genome but instead in a virus particle or in a plasmid: (3) the toxins are synthesized as single chain polypeptides that are only weakly active: (4) the molecules undergo posttranslational processing to give dichain molecules in which a heavy chain is linked to a light chain (molecular weight ratio ~ 2:1) by a disulfide bond, and the dichain molecule is fully active; and (5) the toxins act preferentially on eukaryotic as opposed to prokaryotic cells.

The similarities extend beyond this. For the clostridial neurotoxins and for diphtheria toxin, the tissue-targeting domain appears to reside mainly, though not exclusively, in the carboxy-terminus of the heavy chain.[30-32] For diphtheria toxin, there is compelling evidence that the toxin is internalized by the process of receptor-mediated endocytosis (see Chapter 3). This is followed by a pH-dependent event, during which the toxin or some fragment leaves the endosome (or lysosome) to reach the cytoplasm. It is not yet clear whether the pH-induced event should be characterized as a specific (i.e., formation of a tunnel protein through the membrane) or a nonspecific (i.e., perturbation of the membrane to create avenues for escape) event. However, it is clear that the amino-terminus of the heavy chain of diphtheria toxin possesses the pH-sensitive domain. The evidence for receptor-mediated endocytosis of clostridial neurotoxins is less compelling. However, it has been demonstrated that these toxins have pH-sensitive domains that can create channels in membranes, and these domains are localized in the aminoterminus of the molecules.[33-35]

Diphtheria toxin is known to enter cells, where it acts catalytically to inactive elongation factor 2, a translocase that is involved in protein synthesis.[24] The light chain possesses the enzymatic activity. When the light chain is separated from the heavy chain, it loses its tissue-targeting and channel-forming domains, and thus it is essentially inactive against intact cells. However, when the light chain is introduced into cells by artificial means, it will express its enzymatic effects and poison cells.[20] An enzymatic domain has not been identified in clostridial neurotoxins, although one has been identified in clostridial binary toxins (see below). The enzymatic activity in the latter toxins is found in the light chain.[36] The light chains of these toxins are not very active when added alone to the outside of vulnerable cells, but they are very potent when injected directly into cells.[78] To the extent that the data on diphtheria toxin and the more recent findings on clostridial binary toxins are predictive, they suggest that the light chains of botulinum and tetanus neurotoxins are enzymes, and they exert their effects inside cells.

6.   There are a host of drugs that are known to antagonize the actions of certain agents that must be internalized to exert their effects.[37] Historically these antagonists have been referred to as lysosomotrophic agents, but this term may be misleading. Its origin is related to the fact that these substances are basic, they accumulate in lysosomes, and they counteract the ability of the proton pump to acidify the contents of lysosomes. While these general properties may be true for most of the so-called lysosomotrophic agents, it appears that these drugs are actually heterogeneous and they have several actions.

The most closely studied of the antagonists are ammonium chloride and methylamine hydrochloride, which appear to act similarly, and chloroquine. All three drugs have been shown to antagonize diphtheria toxin.[38,39] Interestingly, they are also very effective antagonists of botulinum neurotoxin.[40,41] The fact that these drugs antagonize botulinum neurotoxin is in itself suggestive evidence that the toxins are internalized. The drugs have been shown to delay the onset of action of many substances that are endocytosed; they have never been shown to alter the effects of a toxin that acts at the cell surface.

7.   The hypothesis that a drug acts at the cell surface to prevent exocytosis seems to imply that a final event in the secretory process has been blocked; the hypothesis that a drug acts inside the cell opens the possibility for many sites of attack. The data now available suggest that the latter is more in accord with the actions of the toxin. Three representative findings illustrate this point.

Dreyer and associates have performed a clever sequence of experiments in which they have poisoned nerve terminals with more than one clostridial toxin.[42] These "double-poisoning" experiments have shown that botulinum neurotoxins type A and type B may act at different steps in the exocytosis of acetylcholine. This type of finding is compatible with the toxins acting inside the cell and at different sites of attack; it is incompatible with the toxins acting on a single final step.

A study by Lupa and Tabti[43] calls into question whether it is the release process per se that is affected. They examined three types of calcium-mediated release processes at the frog neuromuscular junction: facilitation, augmentation, and potentiation. They found that botulinum neurotoxin had a complex effect on facilitation, enhancing the magnitude of the response as well as the rate of decay. The toxin did not change augmentation, but it abolished potentiation. These results appear to rule out any straightforward explanation for toxin action, such as impairing the final step in exocytosis. The results are more in line with the toxin acting inside the cell to modify more than one substrate, and this modification has more than one outcome. Thus, the toxin can have differential effects on three forms of enhanced transmitter output.

Perhaps the most provocative finds are those of Sanchez-Prieto et al.[44] They have studied the effects of various agents and procedures on botulinum neurotoxin type A-induced blockade of glutamate release from cortical synaptosomes. Based on their work with calcium and with the ionophore ionomycin, they proposed that transmitter release hinges on the ability of calcium to cross an intrasynaptosomal membrane or enter an intrasynaptosomal hydrophobic milieu. Furthermore, they hypothesized that it is this intrasynaptosomal barrier, not the plasma membrane, that is the site of toxin action.

Taken collectively, the several observations and arguments presented above weigh in favor of an intracellular site of action. No single observation or argument is compelling, but the amassed findings are very suggestive. They indicate that the toxin penetrates the nerve membrane to exert a paralytic effect. Thus, if the choice is between a putative action on the outside of the cell and a putative action within the cell, the latter is more plausible. But this does not satisfactorily address another possibility. As noted above, the toxin may not completely penetrate the membrane; instead, it may insert into the membrane. What evidence allows one to decide whether insertion into the membrane as opposed to completely crossing the plasma membrane is the more likely alternative?

In a series of electron microscopic autoradiographic studies, Black and associates have provided evidence for internalization of the toxin.[16,29] They purified types A and B neurotoxin to homogeneity and labeled the material with ^{125}I. They confirmed that the heavy chains of the toxins mediated binding, and they demonstrated the existence of a finite number of toxin receptors. Furthermore, they obtained morphological evidence that appeared to support the pharmacological findings alluded to in several of the points above. They demonstrated that there was internalization of bound toxin, and that the process of internalization was inhibited by low temperature and metabolic poisons, but it was enhanced by high rates of nerve stimulation.

It had previously been reported that lysosomotrophic agents could antagonize the onset of toxin-induced neuromuscular blockade, and two possible explanations were proposed.[40,41] The drugs could act at or near the cell surface to inhibit endocytosis (i.e., they act at an early stage in internalization), or they could act in the cell interior to inhibit a processing step required for exiting the endosome (i.e., they act at the final stage of internalization). Interestingly, the morphological data seem to point to both sites of action.[16,28,29] Ammonium chloride and methylamine hydrochloride acted mainly at an early stage in the internalization process, diminishing the proportion of toxin molecules that crossed the membrane. Conversely, chloroquine appeared to act mainly within the cell, perhaps antagonizing a processing step that occurs late during internalization.

The morphological findings seem to reinforce the pharmacological data, and the concordance of findings are supportive of the hypothesis of internalization. The latter point is especially important. The concordance of data lend weight to the idea that the internalized toxin that was monitored was the same toxin that was associated with neuroparalysis. Had there not been concordance, the data might have been interpreted thusly. The toxin binds and then inserts into the membrane, where it exerts its pathological effect. At a later time, the membrane-bound toxin is routed to the lysosome or some other intracellular compartment where it is degraded. The fact that the pharmacological data and morphological data were in accordance with one another renders this speculative explanation unlikely. More accurately, drugs or procedures that enhanced onset of paralysis also enhanced internalization. These drugs and procedures did not enhance the amount of toxin that was membrane associated. The same argument applies for some of the procedures that antagonized poisoning, and most especially for chloroquine.

A particularly notable experiment bearing on the proposed intracellular site of toxin action has recently been published by Penner et al.[23] Their work is a sequel to two earlier manuscripts by Knight, who showed that botulinum neurotoxin could block the release of catecholamines from adrenal medullary cells.[45,46] The earlier paper indicated that only type D neurotoxin was active, but the later paper showed that serotypes A and B could also inhibit exocytosis. The concentrations of toxin and the durations of exposure were much greater than those needed to produce paralysis at the neuromuscular junction. Nevertheless, the general phenomenology of poisoning was similar at adrenal and neuromuscular sites (i.e., neutralized by antitoxin, irreversible blockade, etc.).

One possible explanation for the weak activity of botulinum toxin on medullary cells is that it is not bound or that it does not undergo productive internalization. To circumvent these difficulties, Penner et al.[23] injected the toxin directly into the cytosol of cells. By measuring changes in membrane capacitance, they were able to monitor the process of exocytosis. In the absence of toxin, exocytosis could be triggered by dialyzing a calcium solution into the cell interior. When an identical experiment was done in the presence of botulinum neurotoxin, exocytosis was strongly inhibited.

Findings like these support the notion that botulinum neurotoxin acts in the cell interior, and simultaneously they challenge the idea that the toxin acts at the surface of the plasma membrane, or that it binds and then inserts into the plasma membrane. For the latter to be true, significant amounts of toxin would have to leak from the single cell under study, diffuse throughout the bathing medium, and still be present at sufficiently high concentrations to produce poisoning via the extracellular route. In view of the relative insensitivity of medullary cells,[45,46] this does not seem credible. Alternatively, one would have to argue that there are high affinity toxin binding sites on both the inside and the outside of the plasma membrane. This too is not very credible. The most parsimonious explanation is that the toxin injected directly into the cell interacted with an intracellular substrate to produce blockade of exocytosis.

In summary, there is no finding that proves conclusively that botulinum toxin must be internalized to produce its poisoning effect. Such compelling evidence will not be forthcoming until the substrate for the toxin has been identified and localized. In the absence of definitive evidence, one is obliged to rely on suggestive experiments and on deductions. There are a host of pharmacological and morphological experiments that point to the cell interior as the site of toxin action, and there is a single intracellular injection experiment that reinforces this idea. There is no comparable wealth of data that point to the plasma membrane as the primary site of attack.

III. CLOSTRIDIAL BINARY TOXINS

A. ORIGIN AND SYSTEMIC ACTIONS

Until recently, it was assumed that there were eight different serotypes of botulinum neurotoxin, and these were designated A, B, C_1, C_2, D, E, F, and G.[2,14] However, work done by Sakaguchi and his associates showed that one of these serotypes, C_2, possessed a unique structure. The true neurotoxins are synthesized as single-chain polypeptides having little biological activity. When exposed to certain proteases, the molecules are cleaved to yield a dichain structure in which a heavy chain ($M_r \sim$ 100,000) is linked by a disulfide bond to a light chain ($M_r \sim$ 50,000). It is in this form that botulinum neurotoxin is maximally active.

The type C_2 toxin is fundamentally different. When isolated from the growth medium of bacterial cultures, it exists as two separate and independent polypeptide chains.[47,48] There are no covalent bonds that link the chains, and indeed the two appear to have little or no noncovalent attraction. (It should be noted that no work has yet been done on the relationship between translation and biological activity. No evidence is available to determine whether the two chains are initially synthesized as a single chain polypeptide that might be processed posttranslationally to give two chains). Although the two polypeptides have no obvious attraction to one another, both are needed for maximal expression of biological activity.

When the botulinum C_2 toxin was tested for its ability to block transmitter release, it was found to be essentially devoid of activity.[49] When tested at concentrations one to two orders of magnitude higher than those used with the known neurotoxins, it did not depress transmitter release from peripheral cholinergic or adrenergic nerve endings. The major effect observed *in vivo* was extravasation of fluid, which has been confirmed in *in vitro* and *in situ* experiments.[49-51] A series of experiments *in vivo*, using serial injection of individual chains and antibodies against these chains, suggested that the heavy chain might mediate a binding step and the light chain might be associated with poisoning.[49]

Subsequent *in vitro* work has demonstrated that the heavy chain does play a principal role in tissue binding.[52] In the presence of this component, the light chain will become cell associated; but in the absence of the heavy chain, the light chain has little ability to associate with cells. There appears to be a critically important proteolytic processing step that underlines the interaction between heavy and light chains. In its native state the heavy chain has negligible biological activity, but when the peptide is exposed to trypsin, it is converted to a fully active form. The proteolytic processing step is not needed for the molecule to bind to cells. Instead, it appears to be needed to create an affinity site for the light chain. Unmodified heavy chain does not facilitate the association of light chain with cells, but the trypsin-treated heavy chain does.[53]

When the poisoning due to the binary toxin occurs naturally, the site of action is the gut. Thus, it is one or more cell types in the gut that have receptors for the heavy chain, and the toxic effect of the light chain is expressed inside these cells. Mechanistically, this sequence of events is different from that seen with botulinum and tetanus neurotoxins. For the latter, there is a progression of extracellular trafficking steps that is needed to route the toxin to the target cell. For the binary toxin, the toxin is in immediate juxtaposition to the target cell(s).

B. CELLULAR ACTIONS

The mechanism of action of the light chain has been definitively established.[36] The molecule is an enzyme that possesses ADP-ribosyltransferase activity. G-Actin has been identified as an intracellular substrate that is modified by the toxin, and it may represent the target site of action of the toxin, but additional work is needed to confirm this.[54,55] The toxin ADP-ribosylates proteins in addition to G-actin; the role that these other proteins play in the poisoning of cells has not been determined.

Because of the unique structure-function relationships of this substance, it has been labeled a binary toxin.[19] The individual chains are only weakly active, but the two in combination are very potent. Only a relatively small number of microbial binary toxins have been identified, such as anthrax toxin and leukocidin, and there has been no obvious relationship demonstrated between or among them. However, work on the botulinum binary toxin appears to have been a prelude to the identification of an homologous series of substances. Two additional members of this series have been isolated and characterized.

Clostridium perfringens synthesizes a binary toxin that produces many of the same pathophysiological consequences as the botulinum binary toxin. It too has heavy chain and light polypeptide chains, and the light chain is known to possess ADP-ribosyltransferase activity.[56] *C. spiroforme* also synthesizes a binary toxin whose light chain has ADP-ribosyltransferase activity.[57] All three substances catalytically modify G-actin, though they have affinity for other substrates as well.

Little work has been done to clarify the mechanisms for internalization and trafficking of the clostridial binary toxins. Nevertheless, there are observations that suggest that the toxins must reach the cell interior to express their effects. As just noted, these toxins modify G actin, which is an intracellular substrate. This alone is rather compelling evidence for internalization. Beyond this, experiments have been done with the botulinum binary toxin that not only point to an intracellular site of action but also implicate the light chain as the agent that mediates toxicity.

When the binary toxin is added to the medium surrounding cells in culture, it typically causes profound changes in morphology.[58,59] Cells that are varigated usually contract and eventually assume a spherical shape. Cells that have extensions not only become round, but in addition they often lose their extensions. For example, when the toxin is added to NIE-115 cells, a neuronal cell line that grows neurites or extensions, the elongated portion of the extensions assume a beaded shape and then appear to disintegrate. For cells that are highly vulnerable to the toxin, or that do not tolerate these marked morphological changes, the outcome is lysis. For cells that are more tolerant to the toxin and to the morphological changes, the outcome is a prolonged state of altered cell shape.

A series of experiments have been conducted with cell lines (Y-1; NIE-115) that survive challenge with the toxin.[59] The results show that extracellular application of the toxin causes the expected morphological changes, but equimolar amounts of the heavy chain or the light chain alone do not affect cell shape. Apparently the botulinum binary toxin is like other multiple domain protein toxins in that the holotoxin is needed to exert effects on intact cells. The results are quite different when procedures are used to circumvent the cell membrane. When added to broken cell preparations, or when added to samples of G-actin, the light chain expresses its biological effect. This is not mimicked by the heavy chain, nor does the heavy chain enhance the enzymatic activity of the light chain.

Equivalent results were obtained with intracellular injection of the toxin.[78] Administration of the holotoxin produced the expected effect on the treated cell, whereas all other cells in the culture remained unaffected. When the heavy chain alone was injected into cells, it did not induce morphological changes. However, intracellular injection of the light chain produced effects that were indistinguishable from those induced by intracellular or extracellular application of the holotoxin. These data are strongly indicative that the light chain possesses the active site, and that the poisoning effect is normally expressed within the cell.

Although the results are rather compelling in implicating an intracellular site of action for the binary toxins, there is an absence of data that would clarify issues pertaining to internalization and trafficking. No information is available to indicate whether the toxin enters cells by receptor-mediated endocytosis; experiments to evaluate the potential antagonistic effects of drugs like ammonium chloride, methylamine hydrochloride, and chloroquine have not been reported; there are no data to indicate whether the toxin has a pH-sensitive domain or whether it can be induced to form pores or perturbations in membranes; and there is no evidence to show that the toxin does or does not undergo a procession step before it enters and acts in the cytoplasm. The issue of binding, internalization, and trafficking of binary toxins is clearly one that richly deserves to be explored.

IV. TETANUS TOXIN

A. MICROBIAL ORIGIN

In the context of extracellular and intracellular trafficking, tetanus toxin is surely one of the most remarkable substances known. Current thinking suggests that, at a minimum, the toxin must enter and exit one cell and then enter another cell to express its major pathophysiological effect. Ironically, it is entry and exit from the conduit cell that is somewhat understood, while entry into the target cell is only poorly understood.

Tetanus toxin is synthesized by the organism *C. tetani*, and it is known to exist in only one serotype. The bacterium that makes tetanus toxin is similar to the one that makes botulinum toxin, and the issues that relate toxin production to microbial growth and physiology are largely the same.[60] In particular, the toxin is not known to play any essential role in the economy of the organism.

The genetic material that encodes the toxin has been localized in a plasmid,[61] and recently the entire nucleic acid sequence for the toxin was determined.[62,63] The deduced amino acid sequence for tetanus toxin was compared to the minimal primary structure that has been directly determined for botulinum neurotoxin, and there was evidence for homology. Given these structural data, as well as the known commonalities in origin and biological activity, one can rightly propose that botulinum neurotoxin and tetanus toxin are descendants of the same ancestral parent.

B. MACROSTRUCTURE OF THE MOLECULE

Although the primary structure of the toxin can be deduced from its nucleotide sequence, investigators have not yet been able to link the microstructure of the molecule with its known action. Structure-activity relationships continue to be discussed in terms of the macrostructure of the toxin.

The synthesis, posttranslational processing and experimentally induced modification of tetanus toxin are much like those of botulinum toxin.[60] The molecule is synthesized as a single-chain polypeptide, and estimates of molecular weight from both traditional and molecular biological techniques yield a value of about 150,000. In this form, the molecule is only weakly active. When exposed to trypsin-like enzymes, the toxin is nicked to give a heavy chain ($M_r \sim 100,000$) and a light chain ($M_r \sim 50,000$) linked by a disulfide bond. The N-terminal of the light chain is proline and the C-terminal is alanine; the N-terminal of the heavy chain is serine and the C-terminal is aspartic acid. The two chains are linked by a cysteine that arises in position 467 of the heavy chain.[62,63] The nicked toxin is the biologically active form of the molecule.

When the nicked toxin is exposed to mild papain digestion, it is cleaved to give two, nontoxic polypeptides.[64,65] The larger fragment is composed of the light chain still linked to the amino-terminus of the heavy chain ($M_r \sim 100,000$). The smaller fragment (N-terminal, lysine 865) is the remainder of the heavy chain ($M_r \sim 50,000$). In keeping with the nomenclature suggested above, the light chain and the heavy chain are referred to as such. The amino-terminus of the heavy chain (AA 458 to 864) is labeled H_1, and the carboxy-terminus of the heavy chain (AA 865 to 1315) is labeled H_2. This numbering scheme is adopted from Eisel et al.[62] and it takes into account an initial methionine group that is encoded but then removed during posttranslational processing.

Tetanus toxin mimics botulinum neurotoxin in the sense that the holotoxin is needed to express the characteristic pathophysiological outcome of spastic paralysis. The isolated light chain and isolated heavy chain are not toxic, and the same is true for the isolated H_1 and H_2 fragments.

C. SYSTEMIC ACTIONS

The poisoning due to tetanus toxin exists only as primary infection; no form of primary intoxication has been encountered.[66] In addition, primary infection leading to secondary intoxication occurs only as a variant of wound disease. For example, accidental puncture wounds, nonsterile hypodermic injection, a contaminated umbilicus, etc., can introduce the organism, and this can lead to *in situ* production of toxin. Primary infection leading to secondary intoxication has not been reported as an enteric disorder. There is no evidence that ingested *C. tetani* can colonize the gut and produce toxin locally.

Tetanus toxin produced in the patient's body enters the circulation in the vicinity of the contaminated wound, and it is distributed throughout the periphery.[66] No significant amount of the toxin is capable of penetrating the blood-brain barrier. There appear to be a number of tissues in the periphery to which the toxin binds, but the target organ of major importance is the cholinergic nerve ending of motor fibers.[67] When present at low concentrations, the toxin enters the motor fiber and uses it as a conduit to reach the central nervous system. The toxin exerts an effect in the spinal cord that produces the clinical outcome of spastic paralysis. When present at high concentrations, some fraction of the toxin that enters motor nerve endings remains in the vicinity of the neuromuscular junction, where it produces an effect that resembles that of botulinum neurotoxin. Thus, high concentrations of tetanus toxin are associated with flaccid paralysis. This is, however, an atypical outcome; the prevalent finding clinically is spastic paralysis.

D. PERIPHERAL ACTIONS

A series of pharmacological studies culminated in the hypothesis that botulinum neurotoxin enters peripheral cholinergic nerve endings to exert its effects.[14] An identical paradigm was used with tetanus toxin, and the results were almost identical.[68] The data suggested a three-step sequence, including: (1) binding, (2) movement of the toxin into or through the membrane, and (3) expression of toxicity. There is evidence that these three steps may be associated with structural domains in the toxin molecule.[19,67] The carboxy-terminus of the heavy chain plays a key role in binding, though other portions of the molecule may be involved as well. The amino-terminus of the heavy chain has been shown to possess a hydrophobic domain, and this portion of the molecule can be induced to form channels or membrane perturbations, so it may be implicated in the internalization process. A definitive role for the light chain has not been established, but many assume that it may contain the active site that mediates poisoning.

The state of research that bears on the purported internalization of tetanus toxin is even more tenuous than that for botulinum neurotoxin. The seminal finding for both toxins was the same. At the end of the binding step, the toxins continue to be accessible to neutralizing polyclonal antibodies.[68] However, when tissues are exposed to the proper conditions (e.g., physiological temperature, nerve stimulation), the toxins rapidly disappear from accessibility to antibody. This movement from an antibody-sensitive to an antibody-insensitive site occurs before there is onset of paralysis. Therefore, the assumption is that movement away from the antibody-sensitive site is somehow synonymous with movement toward an active site.

In the section above, a list of experimental observations and deductions that support the hypothesis of botulinum neurotoxin internalization was enumerated. The same approach can be used here, though the list is shorter due to fewer studies that address the issue. The pertinent arguments pertain to: (1) structure-function analogies with diphtheria toxin and the botulinum binary toxin, (2) enhanced activity with enhanced rates of nerve stimulation, (3) the antagonistic effects of certain lysosomotrophic agents, (4) the existence of pH-induced hydrophobic domains and channel formation, and (5) intracellular injection of toxin or fragments from toxin.

The structure-function argument was presented above and need not be belabored. Four potent toxins, botulinum neurotoxin, the botulinum binary toxin, tetanus toxin, and diphtheria toxin have many apparent commonalities in their origins, proteolytic processing, and biological activity. Two of these toxins, the botulinum binary toxin and diphtheria toxin, are known to act on intracellular substrates. Thus, it is tempting to speculate that the other two are also imported to exert their effects.

In addition to the antibody work, there is a neurophysiological observation that may support the hypothesis of internalization. The rate of onset of tetanus toxin-induced neuromuscular blockade is dependent on the rate of nerve stimulation.[68,69] In the absence of nerve stimulation, the neuromuscular blocking properties of the toxin are greatly retarded. As the rate of nerve stimulation increases, there is a proportionate increase in the rate of onset of paralysis. Given the links between exocytosis and endocytosis, one could interpret this work to mean that rapid nerve stimulation promotes endocytosis of toxin.

Tetanus toxin is somewhat like botulinum neurotoxin in its interaction with lysosomotrophic agents; the neuromuscular blocking actions of both are antagonized by ammonium chloride and methylamine hydrochloride.[41] This effect is manifested in two different ways. In the presence of the drugs, tetanus toxin requires a longer time to produce paralysis. In addition, tetanus toxin remains accessible to neutralizing antibody for a longer time than is observed in the absence of drugs. These observations support the notion that ammonium chloride and methylamine hydrochloride are acting relatively early in the paralytic process, perhaps to retard productive internalization. This is in keeping with the fact that the drugs must be added simultaneously with or only shortly after the toxins. If they are added to tissues after poisoning begins, they no longer exert a protective effect. This suggests that the toxin can ultimately "escape" from the antagonistic effects of the drugs. Perhaps when the toxin leaves the endosome or whatever vehicle it uses for internalization, it ceases to be vulnerable to the actions of lysosomotrophic agents.

There is one respect in which tetanus toxin and botulinum neurotoxin differ. The latter is antagonized by chloroquine, but the former is not.[40] One must be cautious about interpreting this result, because chloroquine is itself a neuromuscular blocking agent. The results may point to a true distinction between botulinum neurotoxin and tetanus toxin. Alternatively, chloroquine may be too toxic on neuromuscular junctions to allow for its testing at a concentration that would antagonize tetanus toxin. There are not sufficient data to choose between these possibilities.

There is a parallel line of evidence that may be supportive of that dealing with antibodies and lysosomotrophic agents. Various authors have reported that tetanus toxin will bind to "model" systems, i.e., to neuronal or neuroblastoma tissue. They have further shown that, depending on conditions, the toxin may undergo an internalization step. The principal observation is that toxin, after it is initially bound, can be removed by detergents or proteolytic digestion, but with the passage of time, a large fraction of the toxin moves to a resistant site (for representative studies, see References 70 and 71). These findings could be supportive of the hypothesis that internalization is an integral part of poisoning, but there are reasons to be hesitant. Studies on binding and subsequent detergent or protease treatment are not done in a way that allows one to correlate the reported internalization step with the biological effect of synaptic blockade. This in part is due to the nature of the experiment (e.g., protease itself disrupts tissues), and it is in part due to the model tissues chosen for study (e.g., they do not release transmitter substance). Another problem is that the experimental paradigm does not allow one to distinguish between uptake that is targeted for an active site in the cytoplasm and that which is targeted for degradation in lysosomes. These difficulties hamper interpretation of the work, but the selection of a suitable model tissue could overcome that.

There is a mounting body of evidence that shows that tetanus toxin has a pH-sensitive domain, and the work is analogous to that in which other toxins have been shown to utilize a pH-sensitive domain to cross membranes. Experiments have been done both on intact toxin and on fragments obtained from toxin, and artificial membranes rather than biological membranes have been the focus of research. The findings can be summarized in the following way.[33,34,72]

Tetanus toxin is capable of forming channels in membranes, and this behavior is pH dependent. The portion of the molecule that is involved is the amino-terminus of the heavy chain, and it responds to pH values in the range of 5.0 and lower. In point of fact, it may be that a pH gradient is the critical determinant. The toxin responds when the pH on the proximal side of the membrane is acidic and pH on the distal side is neutral. These pH-induced changes have been shown to be associated with the flux of potassium across membranes, but sizing experiments suggest that much larger molecules could traverse the channels.

Simultaneous with the pH-induced changes that allow the toxin to create pores or perturbations, there are changes in hydrophobicity and in membrane insertion. At low pH, the molecule exposes an otherwise occult hydrophobic domain. Furthermore, experiments with photoactivatable phospholipids show that low pH causes the toxin to insert into membranes.[73] The hydrophobic area would appear to be a characteristic of the conformation of the molecule, because the primary structure does not contain any long stretches of hydrophobic residues.

These findings are compatible with the hypothesis that the toxin undergoes receptor-mediated endocytosis. As the endosome becomes progressively more acidic, due to a proton pump in the membrane, the toxin is induced to insert into the membrane and create channels. This could be a prelude to translocation across membrane. However, the idea of receptor-mediated endocytosis and subsequent acid-induced translocation is still speculative. There exist no compelling ultrastructural studies to confirm the idea. The only study that actually shows endocytosis of tetanus toxin was done on liver cells rather than nerve cells.[74] One cannot know whether endocytosis was a prelude to a biological effect or merely a part of the cell mechanism for engulfing and degrading foreign substances.

Given that so much of the evidence purportedly showing internalization of the toxin is indirect in nature, one might rightly ask whether there is any reason to suspect internalization, aside from arguments by analogy. This issue may have been addressed in a recent study by Penner et al.[23] As discussed earlier, these investigators injected whole toxin or fragments of toxin into individual secretory cells. They were able to show that the intact tetanus toxin molecule was active, and so was a fragment composed of the light chain covalently linked to H_1. The H_2 fragment was not active. Results like these encourage the belief that the toxin is internalized and acts on an intracellular substrate.

E. CENTRAL ACTIONS

Tetanus toxin can be shown to produce peripheral neuromuscular blockade both *in vivo* and *in vitro*, but only when used at relatively high concentrations. When tested *in vivo* at lower concentrations, its principal effect is in the central nervous system, and the result is an increase in efferent motoneuron activity.

The sequence of events that underlie toxin action in the central nervous system can be outlined in the following way.[67,75,76] The molecule initially binds to nerve terminals in the periphery, mainly alpha-motoneurons but also postganglionic and other fibers. The toxin is internalized, after which it is routed up the nerve fiber by retrograde axonal transport. When the toxin enters the cell body, it moves to sites that are postsynaptic and are in close apposition to presynaptic nerve endings. The molecule leaves the cell of origin, crosses the synaptic space, and interacts with nerve endings within the central nervous system. It acts here to block the release of inhibitory transmitters, such as GABA and glycine. By virtue of abolishing inhibition (i.e., producing disinhibition), it releases all volitional and reflex activity in motor fibers. This accounts for the principal neurological disorder in tetanus, which is spastic paralysis.

There is a virtual consensus that this is the scheme underlying toxin action. Uncertainties arise when trying to provide subcellular mechanisms to account for individual steps in the sequence. There is a prevailing assumption that the toxin is endocytosed by peripheral fibers, but in fact there is no direct evidence to show this. The only thing that has been established is that the toxin enters the nerve; the precise mechanism, whether receptor-mediated endocytosis or some other process, is unknown.

There are many substances that are trafficked along the length of the axon, not just tetanus toxin. There does not appear to be a special mechanism for handling the toxin; instead, it probably usurps a normal mechanism that the nerve uses to convey anabolic and catabolic products to and from the cell body.

No one has clarified the means by which the toxin leaves the cell of origin, but there is strong evidence that this must occur before there is onset of poisoning. Erdman and his colleagues have performed an experiment in which toxin was injected peripherally in a muscle, thus giving it access to motor fibers.[77] After a lag time to allow the toxin to be transported into the central nervous system, they injected neutralizing antibody into the intrathecal space. Interestingly, this procedure prevented the onset of paralysis. Presumably the antibody neutralized the toxin as it was leaving the cell of origin and before it interacted with the target cell.

As indicated above, the toxin acts on GABAergic and glycinergic nerve terminals to block transmitter release. This is reminiscent of clostridial toxin action in the periphery, leading to blockade of acetylcholine release. However, it is unclear whether the mechanisms involved are the same. There is as yet no firm evidence that the toxin enters nerve endings in the central nervous system. To be sure, there are several suggestive clues. Some examples are (1) nerves of central origin appear to sequester or endocytose the toxin, (2) the toxin has a pH-sensitive domain that can insert into membranes and form channels, and (3) morphological studies with labeled toxin have given pictures that are compatible with the hypothesis of internalization. Nevertheless, substantial work remains to be done before the issues of internalization and intracellular trafficking can be considered resolved.

ACKNOWLEDGMENTS

The author's research was supported in part by NINCDS grant NS22153 and by DOD contracts DAMD17-85-C-5285 and DAMD17-86-C-6161.

REFERENCES

1. **Gill, D. M.,** Seven toxic peptides that cross cell membranes, in *Bacterial Toxins and Cell Membranes,* J. Jeljaszewicz, J. and Wadstorn, T., Eds., Academic Press, London, 1978, 291.
2. **Sakaguchi, G.,** *Clostridium botulinum* toxins, *Pharmac. Ther.,* 19, 165, 1983.
3. **Gimenez, D. F. and Ciccarelli, A. S.,** Another type of *Clostridium botulinum, Zentralbl. Bacteriol. Parasitenkd. Infektionskr. Hyg. Abt. 1 Orig.,* 215, 221, 1970.
4. **Smith, L. DS.,** *Botulism, The Organism, Its Toxins, The Disease,* Charles C Thomas, Springfield, IL., 1977, 236.
5. **Inoue, K. and Iida, H.,** Conversion of toxigenicity in *Clostridium botulinum* type C, *Jpn. J. Microbiol.,* 14, 87, 1970.
6. **Inoue, K. and Iida, H.,** Phage-conversion toxigenicity in *Clostridium botulinum* types C and D, *Jpn. J. Med. Sci. Biol.,* 24, 53, 1971.
7. **Eklund, M. W., Poysky, F. T., and Reed, S. M.,** Bacteriophage and the toxigenicity of *Clostridium botulinum* type D, *Nature,* 235, 16, 1972.
8. **Eklund, M. W. and Habig, W. H.,** Bacteriophages and plasmids in *Clostridium botulinum* and *Clostridium tetani* and their relationship to production of toxins, in *Botulinum Neurotoxin and Tetanus Toxin,* Simpson, L. L., Ed., Academic Press, New York, in press.
9. **Hall, J. D., McCroskey, L. M., Pincomb, B. J., and Hatheway, C. L.,** Isolation of an organism resembling *Clostridium barati* which produces type F botulinal toxin from an infant with botulism, *J. Clin. Microbiol.,* 21, 654, 1985.

10. **McCroskey, L. M., Hatheway, C. L., Fenicia, L., Pasolini, B., and Aureli, P.**, Characterization of an organism that produces type E botulinal toxin but which resembles *Clostridium butyricum* from the feces of an infant with type E botulism, *J. Clin. Microbiol.*, 23, 201, 1986.

11. **DasGupta, B. R. and Sugiyama, H.**, A common subunit structure in *Clostridium botulinum* type A, B and E toxins, *Biochem. Biophys. Res. Commun.*, 48, 108, 1972.

12. **Wright, G. P.**, The neurotoxins of *Clostridium botulinum* and *Clostridium tetani*, *Pharmacol. Rev.*, 7, 413, 1955.

13. **Gundersen, C. B.**, The effect of botulinum toxin on the synthesis, storage and release of acetylcholine, *Prog. Neurobiol.*, 14, 99, 1980.

14. **Simpson, L. L.**, The origin, structure and pharmacological activity of botulinum toxin, *Pharmacol. Rev.*, 33, 155, 1981.

15. **Habermann, E.**, ^{125}I-Labeled neurotoxin from *Clostridium botulinum* A, Preparation, binding to synaptosomes and ascent to the spinal cord, *Naunyn-Schmiedebergs Arch. Pharmakol.*, 281, 47, 1974.

16. **Black, J. D. and Dolly, J. O.**, Interaction of ^{125}I-labeled botulinum neurotoxins with nerve terminals, I. Ultrastructural autoradiographic localization and quantitation of distinct membrane acceptors for types A and B on motor nerves, *J. Cell Biol.*, 103, 521, 1986.

17. **Silinsky, E. M.**, The biophysical pharmacology of calcium-dependent acetylcholine secretion, *Pharmacol. Rev.*, 37, 81, 1985.

18. **Simpson, L.**, Kinetic studies on the interaction between botulinum toxin type A and the cholinergic neuromuscular junction, *J. Pharmacol. Exp. Ther.*, 212, 16, 1980.

19. **Simpson, L. L.**, Molecular pharmacology of botulinum toxin and tetanus toxin, *Annu. Rev. Pharmacol. Toxicol.*, 26, 427, 1986.

20. **Yamaizumi, M., Mekada, E., Uchida, T., and Okada, Y.**, One molecule of diphtheria toxin fragment A introduced into a cell can kill the cell, *Cell*, 15, 245, 1978.,

21. **Uchida, T., Yamaizumi, M., and Okada, Y.**, Reassembled HVJ (Sendai virus) envelopes containing non-toxic mutant proteins of diphtheria toxin show toxicity to mouse L. cell, *Nature*, 266, 839, 1977.

22. **Yamaizumi, M., Uchida, T., Okada, Y., and Furusawa, M.**, Neutralization of diphtheria toxin in living cells by microinjection of antifragment A contained within resealed erythrocyte ghosts, *Cell*, 13, 227, 1978.

23. **Penner, R. Neher, E., and Dreyer, F.**, Intracellularly injected tetanus toxin inhibits exocytosis in bovine adrenal chromaffin cells, *Nature*, 324, 76, 1986.

24. **Honjo, T., Nishizuka, Y., Kato, I., and Hayaishi, O.**, Adenosine diphosphate ribosylation of aminoacyl transferase II and inhibition of protein synthesis by diphtheria toxin, *J. Biol. Chem.*, 246, 4251, 1971.

25. **Burgen, A. S. V., Dickens, F., and Zatman, L. J.**, The action of botulinum toxin on the neuromuscular junction, *J. Physiol. London*, 109, 10, 1949.

26. **Kozaki, S., Kamata, Y., Takahashi, M., Shimizu, T., and Sakaguchi, G.**, Antibodies against botulinum neurotoxin, in *Botulinum Neurotoxin and Tetanus Toxin*, Simpson, L. L., Ed., Academic Press, New York, in press,

27. **Hughes, R. and Whaler, B. C.**, Influence of nerve-ending activity and of drugs on the rate of paralysis of rat diaphragm preparations by *Clostridium botulinum* type A toxin, *J. Physiol. London*, 160, 221, 1962.

28. **Dolly, J. O., Black, J., Williams, R. S., and Melling, J.**, Acceptors for botulinum neurotoxin reside on motor nerve terminals and mediate its internalization, *Nature*, 307, 457, 1984.

29. **Black, J. D. and Dolly, J. O.**, Interaction of ^{125}I-Labeled botulinum neurotoxins with nerve terminals. II. Autoradiographic evidence for its uptake into motor nerves by acceptor-mediated endocytosis, *J. Cell Biol.*, 103, 535, 1986.

30. **Goldberg, R. L., Costa, T., Habig, W. H., Kohn, L. D., and Hardegree, M. C.**, Characterization of fragment C and tetanus toxin binding to rat brain membranes, *Mol. Pharmacol.*, 20, 565, 1981.

31. **Simpson, L. L.**, The binding fragment from tetanus toxin antagonizes the neuromuscular blocking actions of botulinum toxin, *J. Pharmacol. Exp. Ther.*, 229, 182, 1984.

32. **Bandyopadhyay, S., Clark, A. W., DasGupta, B. R., and Sathyamoorthy, V.**, Role of the heavy and light chains of botulinum neurotoxin in neuromuscular paralysis, *J. Biol. Chem.*, 262, 2660, 1987.

33. **Boquet, P. and Duflot, E.**, Tetanus toxin fragment forms channels in lipid vesicles at low pH, *Proc. Natl. Acad. Sci. U.S.A.*, 79, 7614, 1982.

34. **Hoch, D. H., Romero-Mira, M., Ehrlich, B. E., Finkelstein, A., DasGupta, G. R., and Simpson, L. L.**, Channels formed by botulinum, tetanus and diphtheria toxins in planar lipid bilayers: relevance to translocation of proteins across membranes, *Proc. Natl. Acad. Sci. U.S.A.*, 82, 1692, 1985.

35. **Donovan, J. J. and Middlebrook, J. L.**, Ion-conducting channels produced by botulinum toxin in planar lipid membranes, *Biochemistry*, 25, 2872, 1986.

36. **Simpson, L. L.**, Molecular basis for the pharmacological actions of *Clostridium botulinum* type C_2 toxin, *J. Pharmacol. Exp. Ther.*, 230, 665, 1984.

37. **DeDuve, C., DeBarsy, T., Poole, B., Trouet, A., Tulkens, P., and Van Hoof, F.**, Lysosomotropic agents, *Biochem. Pharmacol.*, 23, 2495, 1974.

38. **Kim, K. and Groman, N. B.,** *In vitro* inhibition of diphtheria toxin action by ammonium salts and amines, *J. Bacteriol.,* 90, 1552, 1965.

39. **Leppla, S. H., Dorland, R. B., and Middlebrook, J. L.,** Inhibition of diphtheria toxin degradation and cytotoxic action by chloroquine, *J. Biol. Chem.,* 255, 2247, 1980.

40. **Simpson, L. L.,** The interaction between aminoquinolines and presynaptically acting neurotoxins, *J. Pharmacol. Exp. Ther.,* 222, 43, 1982.

41. **Simpson, L. L.,** Ammonium chloride and methylamine hydrochloride antagonize clostridial neurotoxins, *J. Pharmacol. Exp Ther.,* 225, 546, 1983.

42. **Gansel, M., Penner, R., and Dreyer, F.,** Distinct sites of action of clostridial neurotoxins revealed by double-poisoning of mouse motor nerve terminals, *Pfluegers Arch.,* 409, 533, 1987.

43. **Lupa, M. T. and Tabti, N.,** Facilitation, augmentation and potentiation of transmitter release at frog neuromuscular junctions poisoned with botulinum toxin, *Pfluegers Arch.,* 406, 636, 1986.

44. **Sanchez-Prieto, J., Sihra, T. S., Evans, D., Ashton, A,, Dolly, J. O., and Nicholls, D. G.,** Botulinum toxin A blocks glutamate exocytosis from guinea-pig cerebral cortical synaptosomes, *Eur. J. Biochem.,* 165, 675, 1987.

45. **Knight, D. E., Tonge, D. A., and Baker, P. F.,** Inhibition of exocytosis in bovine adrenal medullary cells by botulinum toxin type D, *Nature,* 317, 719, 1985.

46. **Knight, D. E.,** Botulinum toxin types A, B, and D inhibit catecholamine secretion from bovine adrenal medullary cells, *FEBS Lett.,* 207, 222, 1986.

47. **Iwasaki, M., Ohishi, I., and Sakaguchi, G.,** Evidence that botulinum C_2 toxin has two dissimilar components, *Infect. Immun.,* 29, 390, 1980.

48. **Ohishi, I., Iwasaki, M., and Sakaguchi, G.,** Purification and characterization of two components of botulinum C_2 toxin, *Infect. Immun.,* 30, 668, 1980.

49. **Simpson, L. L.,.** A comparison of the pharmacological properties of *Clostridium botulinum* type C_1 and type C_2 toxins, *J. Pharmacol. Exp. Ther.,* 223, 695, 1982b.

50. **Jensen, W. I. and Duncan, R. M.,** The susceptibility of the mallard duck (*Anas platyrhynchos*) to *Clostridium botulinum* C_2 toxin, *Jpn. J. Med. Sci. Biol.,* 33, 81, 1980.

51. **Ohishi, I.,** Lethal and vascular permeability activities of botulinum C_2 toxin induced by separate injections of the two toxin compounds, *Infect. Immun.,* 40, 336, 1983.

52. **Ohishi, I. and Miyake, M.,** Binding of the two components of C_2 toxin to epithelial cells and brush borders of mouse intestine, *Infect. Immun.,* 48, 769, 1985.

53. **Ohishi, I.,** Activation of botulinum C_2 toxin by trypsin, *Infect. Immun.,* 55, 1461, 1987.

54. **Aktories, K., Barmann, M., Ohishi, I., Tsuyama, S., Jakobs, K. H., and Habermann, E.,** Botulinum C_2 toxin ADP-ribosylates actin, *Nature,* 322, 390, 1986.

55. **Ohishi, I. and Tsuyama, S.,** ADP-ribosylation of nonmuscle actin with component I of C_2 toxin, *Biochem. Biophys. Res. Commun.,* 136, 802, 1986.

56. **Simpson, L. L., Stiles, B. G., Zepeda, H. H., and Wilins, T. D.,** Molecular basis for the pathological actions of *Clostridium perfingens* iota toxin, *Infect. Immun.,* 55, 118, 1987.

57. **Simpson, L. L., Stiles, B. G., Zepeda, H., and Wilkins, T. D.,** *Clostridium spiroforme* produces an iota-like toxin that possesses monod(ADP-ribosyl) transferase activity, *Infect. Immun.,* 57, 255, 1989.

58. **Ohishi, I., Miyake, M., Ogura, H., and Nakamura, S.,** Cytopathic effect of botulinum C_2 toxin on tissue-cultured cells, *FEMS Microbiol. Lett.,* 23, 281, 1984.

59. **Zepeda, H., Considine, R., Smith, H. L., Sherwin, J. R., Ohishi, I., and Simpson, L. L.,** The actions of the *Clostridium botulinum* binary toxin on the structure and function of Y-1 adrenal cells, *J. Pharmacol. Exp. Ther.,* 246, 1183, 1988.

60. **DasGupta, B. R. and Sugiyama, H.,** Biochemistry and pharmacology of botulinum and tetanus neurotoxins, in *Perspective in Toxinology,* Bernheimer, A. W., Ed., John Wiley & Sons, New York, 1977.

61. **Finn, C. W., Jr., Silver, R. P., Habig, W. H., Hardegree, M. C., Zon, G., and Garon, C. F.,** The structural gene for tetanus neurotoxin is on a plamid, *Science,* 224, 881, 1984.

62. **Eisel, U., Jarausch, W., Goretzki, K., Henschen, A., Engels, J., Weller, U., Hudel, M., Habermann, E., and Niemann, H.,** Tetanus toxin, primary structure, expression in *E. coli,* and homology with botulinum toxins, *EMBO J.,* 5, 2495, 1986.

63. **Fairweather, N. F. and Lyness, V. A.,** The complete nucleotide sequence of tetanus toxin, *Nucleic Acids Res.,* 14, 7809, 1986.

64. **Helting, T. and Zwisler, O.,** Structure of tetanus toxin *J. Biol. Chem.,* 252, 187, 1977.

65. **Helting, T., Zwisler, O., and Wiegandt, H.,** Structure of tetanus toxin, *J. Biol. Chem.,* 252, 194, 1977.

66. **Bleck, T. P.,** Clinical aspects of tetanus, in *Botulinum Neurotoxin and Tetanus Toxin,* Simpson, L. L., Ed., Academic Press, New York, in press.

67. **Habermann, E. and Dreyer, F.,** Clostridial neurotoxins, Handling and action at the cellular and molecular level, *Curr. Topics Microbiol. Immunol.,* 129, 93, 1986.

68. **Schmitt, A., Dreyer, F., and John, C.,** At least three sequential steps are involved in the tetanus toxin-induced block of neuromuscular transmission, *Naunyn-Schmiedebergs Arch. Pharmakol.,* 317, 326, 1981.

69. **Habermann, E., Dreyer, F., and Bigalke, H.,** Tetanus toxin blocks the neuromuscular transmission in vitro like botulinum A toxin, *Naunyn-Schmiederbergs Arch. Pharmakol.*, 311, 33, 1980.
70. **Critchley, D. R., Nelson, P. G., Habig, W. H., and Fishman, P. H.,** Fate of tetanus toxin bound to the surface of primary neurons in culture: evidence for rapid internalization, *J. Cell Biol.*, 100, 1499, 1985.
71. **Staub, G. C., Walton, K. M., Schnaar, R. L., Nichols, T., Baichwal, R., Sandberg, K., and Rodgers, T. B.,** Characterization of the binding and internalization of tetanus toxin in a neuroblastoma hybrid cell line, *J. Neurosci.*, 6, 1443, 1986.
72. **Boquet, P., Duflot, E., and Hauttecoeur, B.,** Low pH induces a hydrophobic domain in the tetanus toxin molecule, *Eur. J. Biochem.*, 144, 339, 1984.
73. **Montecucco, C., Schiavo, G., Brunner, J., Duflot, E., Boquet, P., and Roa, M.,** Tetanus toxin is labeled with photoactivatable phospholipids at low pH, *Biochemistry*, 25, 919, 1986.
74. **Montesano, R., Roth, J., Robert, A., and Orci, L.,** Noncoated membrane invaginations are involved in binding and internalization of cholera and tetanus toxin, *Nature*, 296, 651, 1982.
75. **Bizzini, B.,** Tetanus toxin, *Microbiol. Rev.*, 43, 224, 1979.
76. **Wellhöner, H.-H.,** Tetanus neurotoxin, *Rev. Physiol. Biochem. Pharmacol.*, 93, 1, 1982.
77. **Erdmann, G., Hanauske, A., and Wellhüner, H. H.,** Intraspinal distribution and reaction in the grey matter with tetanus toxin of intracisternally injected anti-tetanus toxoid F (ab)$_2$ fragments, *Brain Res.*, 211, 367, 1981.
78. **Chinn, K. and Simpson, L. L.,** unpublished observations.

Chapter 6

ENDOCYTOSIS AND INTRACELLULAR SORTING OF RICIN

Bo van Deurs, Kirsten Sandvig, Ole W. Petersen, and Sjur Olsnes

TABLE OF CONTENTS

I. INTRODUCTION

Molecules with quite different functions attach to specific binding sites or receptors on the cell surface and are subsequently internalized by receptor-mediated endocytosis (RME). Under normal physiological conditions of cells and tissues, this process serves a variety of functions, such as controlled transport of certain essential substances to the interior of the cell (e.g., cholesterol via low density lipoprotein (LDL); iron via transferrin), removal of altered or harmful molecules from the cells' environment (e.g., alpha-2-macroglobulin protease complexes; asialoglycoproteins), and transport of macromolecules across epithelial barriers (e.g., IgG; polymeric IgA). Furthermore, following ligand (e.g., insulin; epidermal growth factor [EGF]) binding, ligand-induced transport of receptors to the interior of the cell may result in down-regulation of the corresponding receptors.[1,2]

The endocytic pathway to the interior of the cell is not protected against opportunistic invasion of certain dangerous molecules. Thus, many kinds of enveloped and naked viruses,[3-6] as well as plant and bacterial toxins with intracellular sites of action,[7-9] are internalized by the cell after binding to specific components on the cell surface, thereby imitating RME.

The toxic plant protein ricin has been subjected to detailed studies using various biochemical and morphological approaches.[7-16] Binding, uptake, and intracellular trafficking of ricin are studied mainly for two reasons. (1) Ricin serves as a valuable tool for elucidating intracellular pathways of internalized ligands and membrane molecules in general, and also represents, together with several other toxins, an interesting model for studying posttranslational translocation of proteins across membranes.[17] (2) Ricin is frequently used in the construction of immunotoxins designed to kill specific target cells, such as cancer cells with the purpose of developing potential anticancer drugs.[18-21] It is, therefore, of crucial importance that detailed information on the intracellular trafficking of ricin is obtained and that the intracellular site(s) of ricin A chain translocation to the cytosol is defined.

Biochemical data on ricin binding, internalization, and intracellular action have been discussed in recent reviews.[7-9] Also, many reviews have dealt with the structural organization of intracellular compartments involved in the sorting and processing of various endocytosed ligands and receptors.[1,2,22-27] However, much less structural information on ricin binding, uptake, and intracellular routing is available. The purpose of this review is to outline current knowledge on endocytic pathways of ricin, with emphasis on recent electron microscopic data, and to discuss some implications and problems related to this topic.

II. RICIN BIOCHEMISTRY

Ricin is synthesized as a single polypeptide in maturing castor beans, where it accumulates in the storage granules of the seeds.[28] After synthesis, the toxin is rapidly cleaved into two S-S linked polypeptides, termed the A- and the B-chain.[7,29,30] The molecular weight of the two chains is 30,625 and 31,431 kDa, respectively. The B-chain contains two galactose-binding sites with somewhat different affinity for lactose.[31] These sites are responsible for the binding of the toxin to carbohydrates at the cell surface, a process that is required for toxic effect on intact cells.

Both of the constituent polypeptide chains in ricin are glycoproteins.[7] The major part of the carbohydrate is carried by the B-chain. These carbohydrates may play a role in binding of the toxin to certain cells containing mannose receptors (see Section III).

The A-chain has enzyme activity and carries out the function that eventually leads to cell death. This chain is translocated to the cytosol where it inactivates in an enzymatic manner the 60S ribosomal subunits. Recent work by Endo et al.[32,33] has revealed that ricin A chain is a highly specific N-glycosidase that removes adenine from a single adenosine

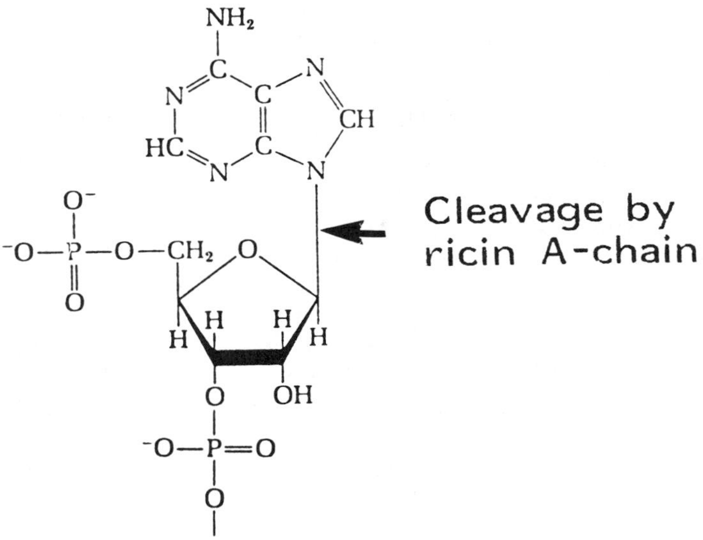

FIGURE 1. Enzymatic action of ricin A-chain. The A-chain is a specific N-glycosidase that removes adenine from a single adenosine residue (A4324) located in a highly conserved region near the 3'-end of 28-S RNA.

residue in 28S RNA. This residue is located near the 3' end of the RNA. The phosphoribose backbone is not cleaved in the process (Figure 1).

It is interesting that a number of other plant toxins, such as abrin and modeccin, have been found to have the same enzymatic effect as ricin.[32,33] Surprisingly, a bacterial toxin, shigella toxin also has been found to have the same enzymatic effect as ricin. In addition the A chain of shigella toxin has considerable sequence homology to ricin A-chain.[34,35]

III. RICIN BINDING TO THE CELL SURFACE

Ricin binds to a large number of sites at the cell surface which contain terminal galactose residues. Different cell types bind between 10^6 and 10^8 toxin molecules per cell.[36-38] The binding sites are partly carried by glycoproteins and partly by glycolipids.[39] Labeling with ricin is, therfore, a general method of labeling the cell surface membrane.

The toxin has two binding sites for galactose.[31] Because these have somewhat different affinity for the sugar, and as a consequence, probably also different affinity for cell surface carbohydrates, it is not clear whether both sites, or only the high-affinity site, are involved in cell binding. The weak agglutinating effect of pure ricin could be due to traces of aggregates formed. The fact that tyrosine modification by N-acetyl-imidazole treatment strongly inhibits the binding[40] in spite of the fact that tyrosine is involved in galactose binding at only one of the sites,[41] indicates that only one site, and presumably the high-affinity site, binds strongly to cells. Also the fact that the toxin is inefficient in agglutinating cells indicates that only one site binds strongly to cells.

Ricin may also be bound to certain cells by an entirely different mechansim. As mentioned above (Section II.), ricin is a glycoprotein and contains mannose-rich carbohydrates linked to both of its constituent polypeptide chains. On the surface of cells of the reticuloendothelial system, mannose receptors are present that may bind these carbohydrates.[2] It has been shown that binding by this mechanism results in efficient intoxication of cells.[42-43] On the other hand, deglycosylation of the toxin abolishes binding and intoxication by this pathway.[44]

To study ricin binding, as well as ricin uptake (see Sections IV and V), at the ultrastructural level, ligand conjugates or immunocytochemical detection of the ligand can be used. In both cases, the probe by which ricin or the antibody against ricin is visualized in the electron microscope can be either enzymatic, i.e., horseradish peroxidase (HRP), or particulate (ferritin or colloidal gold).

Both monovalent and polyvalent conjugates of ricin and HRP (ricin:HRP 1:1, and 2 or more:1) prepared by the SPDP (*N*-succinimidyl-3-[2-pyridylthio]-propionate) method[45,46] followed by gel filtration[10] bind evenly to the cell surface at 4°C. When cells are labeled with ricin-HRP at this temperature and then washed with buffer and further incubated for various periods of time at 37°C, ricin-HRP is still evenly distributed on the cell surface, including microvilli and various coated and uncoated cell surface pits (see Section IV). Similar results are obtained using immunoperoxidase cytochemistry to detect ricin on the cell surface. This indicates that ricin binding sites are distributed all over the cell surface and that they are internalized only slowly. Possibly, some binding sites are not internalized at all.

In contrast, when transferrin receptors are visualized at the cell surface either by transferrin-HRP binding or by incubating cells with antitransferrin receptor antibody followed by HRP-conjugated secondary antibody, the labeling is preferentially seen within or close to coated pits,[16] indicating that these receptors are associated with coated pits even without ligand binding. Alternatively, when EGF receptors are tagged with EGF-HRP at 4°C, they appear to be evenly distributed on the cell surface. However, following incubation for some minutes at 37°C, the labeling largely disappears from the cell surface due to internalization.[47] We have seen a similar pattern when cells were incubated with shigella toxin-HRP at 4°C and after 15 to 60 min at 37°C.[137]

By using a ricin-gold conjugate, an even surface labeling was also obtained at 4°C[12] (Figure 2). Following warming of the prelabeled and washed cells to 37°C, the number of gold particles per micron cell surface decreased due to endocytosis. Moreover, the ricin-gold still present on the cell surface (after 30 min at 37° about 45%) often appeared in more or less aggregated patches. This phenomenon is most likely due to the fact that ricin-gold is a polyvalent ligand complex which may induce receptor redistribution and aggregation upon binding at 37°C. Thus, when unconjugated ricin is bound to cells at 4°C and the cells are then incubated at 37°C, or when cells are incubated at 37°C with ricin present in the medium, subsequent analysis of ultracryosections treated with antiricin followed by protein A-gold (PAG) never reveals marked patching. By the latter technique, ricin appeared relatively evenly distributed over the cell surface[15] (Figure 2). The reason for the apparent discrepancy in surface labeling at 37°C with ricin-gold and polyvalent ricin-HRP remains unclear.

The binding specificity of the ricin-HRP and ricin-gold conjugates described above can be checked by preincubation of the cells either with an excess of unlabeled ricin or with 0.1 *M* lactose. In biochemical experiments designed to measure the amount of endocytosed ricin under various conditions, cell surface-associated ricin is removed from the cells with lactose. We found that only monovalent ricin-HRP could be removed in this way. Polyvalent ricin-HRP as well as ricin-gold could not be removed with lactose.[12,13]

Altogether, ultrastructural studies using various approaches indicate that ricin binding sites are evenly distributed over the cell surface. Moreover, even though bound toxin is internalized at 37°C, the cell surface never becomes depleted of binding sites (at least within the period of time available for experiments, i.e., the time until the cells are intoxicated). A reason for this is that the overall internalization of ricin binding sites is a slow process. Possibly, some binding sites are never internalized. Also, some of the toxin is rapidly recycled to the cell surface.[48] From a technical point of view, monovalent ricin conjugates should be preferred to polyvalent ones to obtain a precise, undisturbed localization of the binding sites

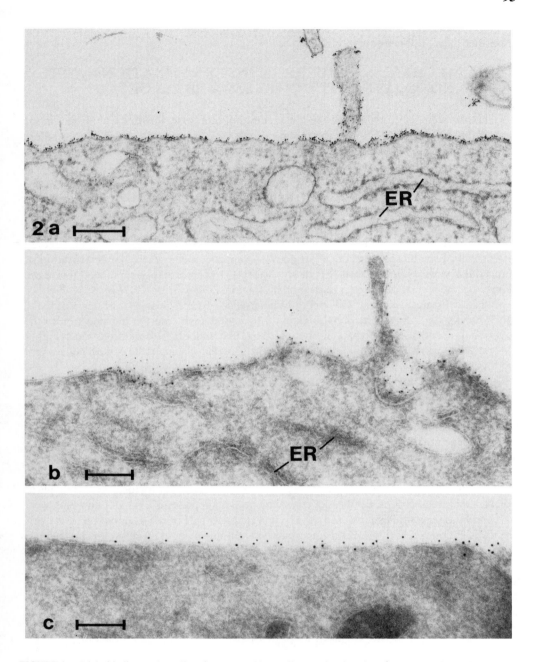

FIGURE 2.   Ricin binding to the cell surface. In (a) Vero cells were incubated at 4°C with a ricin-gold conjugate (gold size about 5 nm). In (b) BHK-21 cells were incubated at 4°C with unconjugated ricin, fixed and frozen. Ultracryosections were then cut and incubated with antiricin antibody followed by protein A-gold (PAG; 8 nm). In (c) BHK-21 cells were incubated with ricin for 60 min at 37°C and then processed as in (b) except that a 10-nm PAG was used. It is evident that ricin binding sites are numerous and evenly distributed on the cell surface, also when ligand binding (unconjugated ricin) occurs at 37°C. ER, endoplasmic reticulum; bars, 0.25 μm.

on the cell surface. Furthermore, the binding of monovalent conjugates is specific in the sense that they can be removed with lactose.

IV. MECHANISM(S) OF RICIN INTERNALIZATION—ARE NONCOATED PITS AND VESICLES INVOLVED?

The involvement of endocytosis in the penetration of a toxin was first established in the case of diphtheria toxin.[49,50] This toxin requires transport to an acidic compartment before translocation to the cytosol can take place. Since the required pH (<5.3) is found only in intracellular vesicles or vacuoles, such as endosomes and lysosomes,[55] it follows that transport to one of these compartments must precede translocation across the membrane. It has later been established that the translocation takes place from the endosomes.[52] Also a number of other toxins (e.g., modeccin and *Pseudomonas aeruginosa* exotoxin A) require low pH at some stage during their entry.[53,54]

In the case of ricin and certain related toxins (abrin and viscumin), low pH is not required for entry. In fact, these toxins are most active under conditions where the acidification of intracellular vesicles is inhibited.[55,56] In spite of this, evidence has accumulated that these toxins also are translocated from intracellular structures.

When cells are incubated with ricin in the absence of Ca^{2+} or in the presence of Co^{2+} which blocks Ca^{2+} transport, ricin is taken up by endocytosis, but the translocation to the cytosol is prevented.[56,57] The data in Figure 3 show that under these conditions much higher concentrations of ricin are required to inhibit protein synthesis than in the control cells where Ca^{2+} is present.

If the cells are then washed and treated with antiricin to inactivate any ricin present at the cell surface and subsequently incubated overnight in the presence of Ca^{2+} and antibody, the cells are intoxicated to the same extent as if the whole experiment had been carried out in the presence of Ca^{2+}.[56] This demonstrates that ricin endocytosed under conditions where it cannot penetrate to the cytosol is capable of intoxicating the cells without returning to the cell surface (where it would be inactivated by antibodies). Furthermore, the finding that there is no detectable difference in the extent of intoxication whether or not exposure to the toxin occurs in the presence of Ca^{2+} where translocation to the cytosol can take place indicates that translocation from endocytic compartments represents a major entry pathway.

Morphological studies have also clearly revealed that part of the ricin bound to the cell surface is endocytosed.[12-16,58,59] An important and still unsolved question in this connection is *how* ricin is internalized.[12,16]

It is well established that membrane-enveloped viruses are taken up by coated pits.[3,5,6] In a study on the uptake of Semliki forest virus (SFV), Marsh and Helenius[6] calculated that the coated pits and vesicles involved in the uptake of this virus also may account for the nonspecific fluid phase endocytosis found in the BHK cells used. This study could, therefore, be taken to indicate that the coated-pit pathway is the only endocytic pathway, and this view has actually been the predominant one in many laboratories. However, it should be stressed that while the evidence for uptake of numerous molecules — in addition to viruses — via the coated-pit pathway is overwhelming and has been the topic of several reviews,[1,22,60,61] this does not prove that another, alternative, endocytic mechanism does not exist as well. The observation of ricin in coated pits at the surface of various cell types[12,16] (Figure 4) strongly suggests that at least some ricin is taken up via these structures, but it does not exclude the possibility of some ricin being taken up by other means.

Many independent pieces of evidence suggest that smooth or uncoated areas at the cell surface (i.e., areas not showing the characteristic clathrin coat of coated pits and vesicles in conventional sections for electron microscopy) are also involved in endocytosis. For example, it was reported that HLA antigens tagged by IgG-ferritin aggregated in smooth

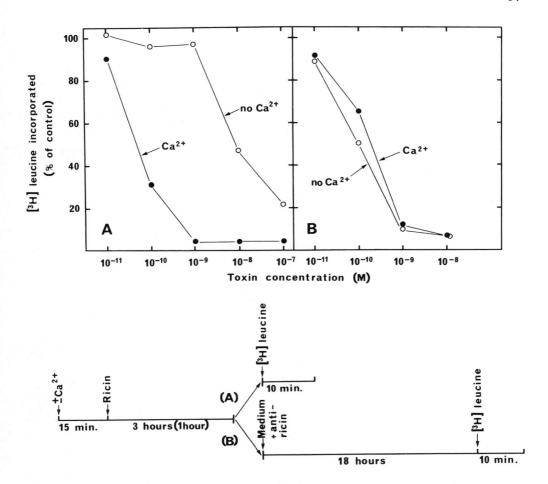

FIGURE 3.   Ability of ricin endocytosed in the absence of Ca^{2+} to intoxicate cells. Vero cells were incubated with increasing concentrations of ricin in a buffer containing either 2 mM $CaCl_2$ or 2 mM $CoCl_2$. In (A) protein synthesis was measured after 3 h of incubation; in (B) the cells were transferred to a medium containing $CaCl_2$ and neutralizing amounts of antiricin antibody after 1 h, and the cells were then incubated overnight before protein synthesis was measured.

pits rather than in coated pits of cultured human fibroblasts and that the conjugate was subsequently internalized.[62] Colloidal gold conjugates of tetanus and cholera toxins, which bind to membrane lipid receptors, were similarly reported to be endocytosed via smooth pits and vesicles in cultured liver cells.[63] EGF has been reported to induce formation of smooth pits upon binding to its receptor on cultured epitheloid A 431 cells. These smooth pits are thought to be responsible for the subsequent uptake of the ligand-receptor complex.[64] In endothelial cells, smooth pits and vesicles have been ascribed to receptor-mediated transcytosis of albumin-colloidal gold.[65]

Against the conclusion of these studies, it may be argued that the finding of a ligand-marker complex in a smooth pit at the cell surface does not necessarily mean that the pit is involved in endocytosis. Furthermore, ligand-marker complexes found in intracellular compartments may have been rapidly taken up by coated pits and vesicles. Indeed, in time sequence studies, very short time points (i.e., 0 to 5 min) and quantitation must be included to support the conclusion that the ligand-marker conjugate does not occur in coated pits to any significant degree before internalization and intracellular sequestration take place. However, if coated and smoth endocytic pits pinch off at different rates, this would make the usefulness of such quantitation questionable.

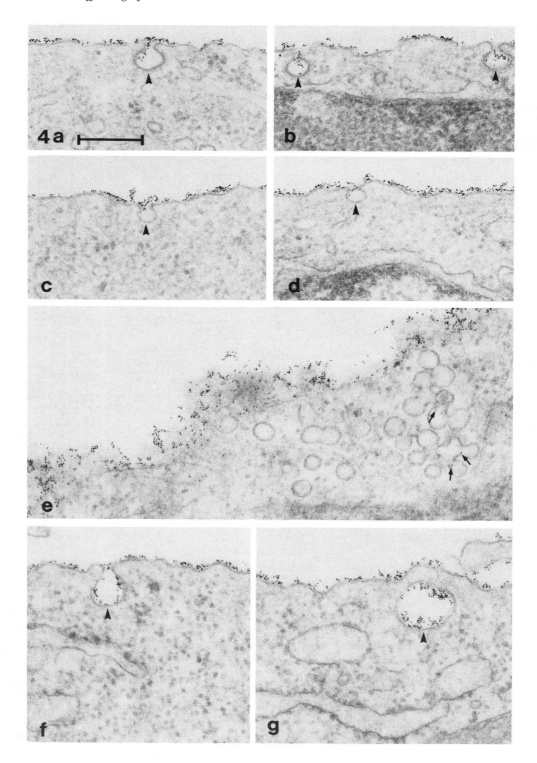

FIGURE 4.

Potassium depletion in combination with hypotonic shock is known to deplete the surface of some cell types of coated pits, thereby abolishing the uptake of, for instance, LDL receptors.[66-68] Using this approach, Moya et al.[69] and Madshus et al.[4] found that when Hep-2 cells were exposed to potassium depletion in combination with a brief hypotonic shock, virtually no coated pits were present at the cell surface and the uptake of transferrin was almost completely blocked, whereas the uptake of ricin was only slightly reduced (see also Reference 70). Madshus et al.[4] also reported that under these conditions the cells were strongly protected against poliovirus while the cytopathogenic effect of human rhinovirus type 2 (HRV 2) and encephalomyocarditis virus (EMC-virus) was not inhibited. Since productive entry of HRV 2 requires transfer to an acidified intracellular compartment, these observations suggested that HRV 2, as well as ricin, are taken up by a pathway not involving coated pits and vesicles.

Recently, we have developed another method of selectively inhibiting endocytosis via coated pits. We showed that acidification of the cytosol of various cell lines inhibited endocytosis of transferrin and EGF.[16] The acidification was carried out in three different ways: (1) incubation of cells with NH_4Cl followed by incubation in Na^+-free medium; (2) incubation of cells with acetic acid; and (3) incubation with isotonic KCl containing nigericin and valinomycin.

Using quantitative electron microscopy, we found that the number of coated pits at the cell surface was similar in acidified cells and control cells. Moreover, by using a monovalent transferrin-HRP conjugate (transferrin-HRP ratio approximately 1:1) and a monoclonal antihuman transferrin receptor antibody followed by a secondary antibody conjugated to HRP, we showed that transferrin receptors were present in the coated pits of both control and experimental cells. The most likely explanation of the abolished uptake of transferrin under these experimental conditions is that acidification has prevented the formation of coated endocytic vesicles from coated pits, possibly by some structural alteration of the clathrin cage that was not apparent in conventional sections for electron microscopy. Thus, the uptake of molecules, such as transferrin, that depend on binding to receptors prelocated or predominantly located in coated pits is also prevented. However, if certain molecules in part or completely bind to surface "receptors" which are not localized in coated pits, they might still be taken up. In fact, acidified cells showed only a slight reduction in the uptake of ricin and the fluid-phase marker Lucifer Yellow (the reduction depending on the cell type tested)[16] (Table 1). Accordingly, also the acidification experiments support the notion that an alternative and by definition not clathrin-coated endocytic pathway does exist. This pathway may actually be responsible for the major part of ricin uptake in some cell types.

Evidence for a partial inhibition of a noncoated endocytic pathway also has been reported. Thus, studies by Li et al.[71,72] showed that mutants of LM fibroblasts with decreased ability to undergo polyethylene glycol-induced cell-to-cell fusion exhibit reduced fluid-phase and nonspecific adsorptive endocytosis, whereas RME of transferrin is normal. These observations suggest that two endocytic mechanisms exist, each of which has different requirements for vesicle formation.

Recently it was shown that introduction of anticlathrin antibodies into the cytosol caused aggregation of clathrin within the cytoplasm, a marked reduction in the number of coated pits at the cell surface, and a deformation of the remaining coated pits. However, uptake of SFV and the fluid-phase marker Lucifer Yellow was decreased by only 40 to 50%.[73]

FIGURE 4.   Possible endocytic structures at the cell surface, as visualized after incubation of Vero cells at 4°C with a ricin-gold (5 nm) conjugate: (a) and (b) show some coated pits containing ricin (ricin conjugate) and (c) and (d) some smooth pits (arrowheads); (e) is a tangential surface section showing how the small smooth pits can appear in a rather complex, interconnected arrangement which does not makes an endocytic function obvious. Note the presence of a few ricin-gold particles (arrows) within the pits. (f) and (g) show examples of larger, smooth pits with numerous ricin-gold particles (arrowheads). Note that the indicated vesicular profiles in (g) must be surface connected since the ricin-gold incubation was performed at 4°C. Bar, 0.25 μm.

TABLE 1
Effect of Acidification of the Cytosol on Endocytic Uptake of Ricin and Lucifer Yellow[a]

Acidification method	Cell line	% of control	
		Endocytic uptake of ^{125}I-ricin	Endocytic uptake of Lucifer Yellow
NH₄⁺-prepulse	Vero	82 ± 15	84 ± 12
NH₄⁺-prepulse	A 431	79 ± 10	
NH₄⁺-prepulse	Hep 2	69 ± 17	
NH₄⁺-prepulse	MCF 7	65 ± 10	
Acetic acid	Vero	78 ± 8	

[a] Data are from Reference 16, where the methods of acidification of the cytosol are also described.

Although other explanations may be given, the observations are not incompatible with the concept of an alternative endocytic pathway, not involving clathrin.[73]

A very important question arises concerning the structural equivalent to this, still somewhat hypothetical, alternative endocytic pathway. The only feasible approach so far has been to look for surface structures (pits) binding ricin (visualized as ricin conjugates or by immunocytochemistry) but not transferrin (visualized as transferrin conjugates).[16] While the majority of the cell surface transferrin receptors are localized within or close to coated pits,[16,64,74,75] and are absent from, or rare in various smooth pits, ricin is frequently localized in such smooth structures.[12,16]

In general, smooth pits appear more frequently (2 to 10 times; cf. Reference 4) at the cell surface than coated pits, but they are for obvious reasons difficult to define and quantify.[4] Most important, they lack a "marker", such as the clathrin coat of coated pits, which makes the latter structure identifiable when it is only slightly invaginated, or even when it does not appear invaginated at all. Specific markers of smooth pits have not yet been identified.

In particular, a small version of the smooth surface pits (diameter about 50 to 70 nm) is very abundant in many cell types (Figure 4) and, as discussed at the beginning of this chapter, has often been considered endocytic. However, at least as far as endothelial cells are concerned, extensive thin serial section analysis of tissue prepared by conventional electron microscopic techniques, as well as by rapid freezing-freeze substitution, has revealed that the smooth vesicular profiles seen in random sections are almost all surface connected. It is, therefore, unlikely that they are endocytic and involved in transcytosis in endothelial cells.[76-78]

Also, some small smooth pits at the cell surface may represent recycling vesicles carrying membrane from intracellular endocytic compartments back to the cell surface (see Section V).[24,79-81] Other smooth pits at the cell surface may represent secretory vesicles coming from the Golgi complex. The morphology of such vesicles is largely unknown. Small, smooth pits at the cell surface could also be permanent surface-associated structures, with unknown functions, as mentioned above for endothelial cells. This notion may apply in particular to the numerous caveolae seen in fat cells and smooth muscle cells.[82,83] Therefore, it is an open question which part of the total population of small, smooth pits seen at the cell surface of many cell types it is that plays a role in the alternative endocytic pathway.

A possible alternative to the small, smooth pits at the cell surface as candidates for the alternative endocytic pathway would be some larger, uncoated invaginations (Figure 4).[12,16] These invaginations have a diameter of up to about 0.4 μm, and whereas they were not found to contain transferrin or transferrin receptors in Hep-2 cells, they contained ricin (ricin-gold).[16] The invaginations can be observed as apparently free vacuoles containing membrane-

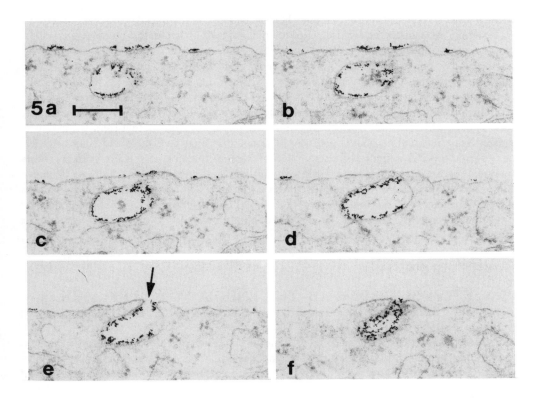

FIGURE 5. Vero cells incubated with ricin-gold (5 nm) for 20 min at 37°C. Six very thin (about 20 nm) consecutive sections (a to f) of a vesicular structure containing large amounts of ricin-gold are shown. Note that although the structure may be taken as a peripheral endosome in (b), it clearly communicates with the cell surface in (e) (arrow). Also note the uneven or patchy surface distribution of ricin-gold at 37°C; while much gold is seen on the surface in (a), no gold is seen in (f). Bar, 0.25 μm.

bound ricin in the peripheral cytoplasm of cells incubated at 4°C (Figure 4). Therefore, the vacuoles must communicate with the cell surface.[12,16] Also, apparently free, ricin-containing peripheral vacuoles in cells that have been incubated with the toxin at 37°C can be shown by serial sectioning to communicate with the cell surface (Figure 5).[16] These observations suggest that some structures with the appearance of peripheral endosomes are actually cell surface invaginations. Whether they represent endocytic pits in the process of pinching off remains an open question. However, it is obvious that if they are involved in endocytosis, even a relatively few such large surface pits can be of quantitatively great importance in uptake of ricin and fluid-phase markers.

The discussion of whether only (clathrin-) coated pits and vesicles are involved in endocytosis or whether other structures also play a role is further complicated by the observation that some coat material seen on the cytoplasmic surface of endocytic structures may not be clathrin. Thus Brown et al.[84] recently described endocytosis of HRP by some "coated" tubulovesicular structures in intercalated cells of the kidney collecting duct. In freeze-fracture, the membrane of these coated structures showed characteristic intramembrane particles on the P-face, and a distinct cytoplasmic coat was revealed by conventional thin-section electron microscopy. Immunocytochemistry showed that the coat of these specialized endocytic structures did *not* contain clathrin.[85] It will, therefore, be interesting to see in other cell types whether endocytic "coated" pits and vesicles generally assumed to represent typical clathrin-coated structures are in fact coated by material different from clathrin.

Another complication in this discussion would be if "clathrin-coated pits" at the cell surface should represent more than one (homogeneous) population of endocytic structures, for example with different requirements for pinching-off,[16] but at the moment this aspect remains purely speculative.

In conclusion, it is clear that at least some ricin molecules bound to the cell surface are taken up via coated pits and vesicles (Figure 4). Thus ricin in part imitates physiologically important ligands such as transferrin and LDL, which enter cells via glycoprotein receptors with affinity for coated pits. Moreover, several independent lines of experimental evidence (e.g., potassium depletion and acidification of the cytosol) suggest that ricin is also internalized via a nonclathrin-coated pathway. The main problem in establishing such an alternative pathway and defining its structural equivalent is the lack of specific markers. Only by obtaining markers, such as antibodies against cell surface molecules which are excluded from coated pits and are prelocated (at 0 to 4°C) in, and selectively internalized by smooth pits, can this controversial issue be resolved. The receptors which are internalized via coated pits are glycoproteins. However, tetanus and cholera toxins bind to glycolipid receptors and appear to be internalized via smooth pits; ricin also binds, in part, to glycolipids. Therefore, cell surface glycolipids appear to be a promising possibility in future studies which search for markers for smooth pits.

V. INVOLVEMENT OF THE ENDOSOMAL SYSTEM IN INTRACELLULAR TRAFFICKING OF RICIN

When labeled ricin is bound to the cell surface and the cells are subsequently incubated at 37°C, part of the bound toxin is taken up by endocytosis. This can be measured conveniently due to the fact that toxin present at the cell surface is rapidly released by treating the cell with lactose[48] which is a competitive inhibitor for the binding (see also Section III). Toxin that has been taken up by endocytosis is not released under these conditions. Using this method, we found that about 15% of the cell-associated toxin is endocytosed in 10 min (Figure 6) (see also references 37 and 56).

Many physiological ligands that are taken up by endocytosis from coated pits are internalized at a much higher rate. Thus, as much as 60% of the cell-associated transferrin was taken up by 10 min (in Vero cells) (Figure 6) in agreement with data obtained in other cells.[86,87] The reason for the comparatively slow overall endocytosis of ricin is presumably that it binds to a large variety of different receptors, many of which may be internalized slowly or represent essentially stationary surface structures.

Internalization of ricin has also been documented by means of morphological techniques. Ricin has been detected intracellularly in endosomes, using both preembedding immuno-peroxidase cytochemistry[13] and postembedding immunogold cytochemistry[15] (Figure 7). Moreover, various ricin conjugates (ricin-HRP; ricin-gold) (Figure 7) have been traced to endosomes after only a few minutes of exposure to the cells.[12-14] As for other internalized ligands, endosomes represent the first intracellular station for endocytosed ricin, no matter whether it is taken up by coated pits, by smooth pits, or by both (see Section IV).

Endosomes represent an acidified compartment[51] where the low pH (between 5 and 6.5) is responsible for the dissociation of some ligands (e.g., LDL) and their receptors, a prerequisite for further sorting and trafficking of the ligand to lysosomes and the receptor back to the cell surface.[1,2,51] In contrast, the affinity of apo-transferrin (i.e., transferrin after low pH-mediated release of iron) for its receptor is not reduced by low pH in endosomes. Therefore, apo-transferrin is recycled from endosomes to the cell surface bound to its receptor. However, at the neutral pH at the cell surface, apo-transferrin dissociates and remains free until it has again bound iron. A third example is the receptor for the Fc fragment of the IgG molecule. Without ligand binding, this receptor cycles between the cell surface and endosomes in macrophages. Following ligand binding, however, ligands and receptors

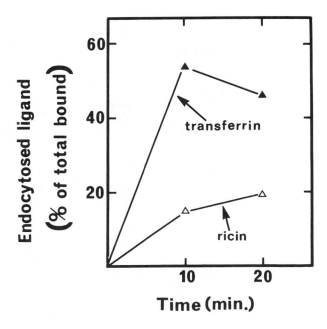

FIGURE 6. Endocytic uptake of transferrin and ricin in Vero cells. Vero cells were incubated with ^{125}I-ricin or ^{125}I-transferrin at 37°C, and the amount of cell-associated ligand as well as the amount of internalized ligand was measured after increasing periods of time.

do not dissociate at the low endosomal pH. Therefore, both ligands (for instance, an IgG-coated particle) and receptors are eventually transported to lysosomes, where they are degraded. In this case, the receptors are said to be down-regulated upon ligand binding.[1,2,51]

Internalized ricin also remains attached to its binding sites at the low pH in endosomes. Thus, only about 22% of the ricin initially bound at pH 7 was found to be released at pH 5.0.[15] It is known from biochemical studies that internalized ricin molecules are recycled to the cell surface, probably attached to their binding sites.[48] Thus we found that the majority of internalized toxin later released from cells was released as intact toxin.[37,48] The kinetics of release showed that a fraction of the internalized toxin returned rapidly (within 30 min) to the cell surface, whereas another fraction returned at a much slower rate.

The release of the toxin from the cells also showed interesting temperature kinetics. Thus, Arrhenius plots showed biphasic curves with an increase in exocytic rate at about 20°C. Below this temperature, there was no measurable degradation of the toxin by the cells.[48]

However, direct morphological evidence for this recycling does not exist. Based on studies on endocytosis of other ligands,[88,89] it is suggested that ricin molecules and their binding sites are recycled to the cell surface from tubular portions of the endosomal system (CURL) (Figure 7). Experiments with HRP and cationized ferritin and quantitative electron microscopical analysis have indicated that the presumptive recycling vehicle connecting the endosomal system and the plasma membrane could be small (diameter 50 to 100 nm) uncoated vesicles or tubules.[79-81]

Sorting of ligands and receptor traffic in the endosomal system also may depend on the valency of the ligand (i.e., the number of binding sites).[13,51,90-94] Evidence for this comes from studies of the Fc receptor system.[91-93] Monovalent ligands bound to this receptor allow continuous recycling in an "undisturbed" way, whereas polyvalent ligands (for instance, particles coated with IgG) change the intracellular routing of the receptor. It has been reported

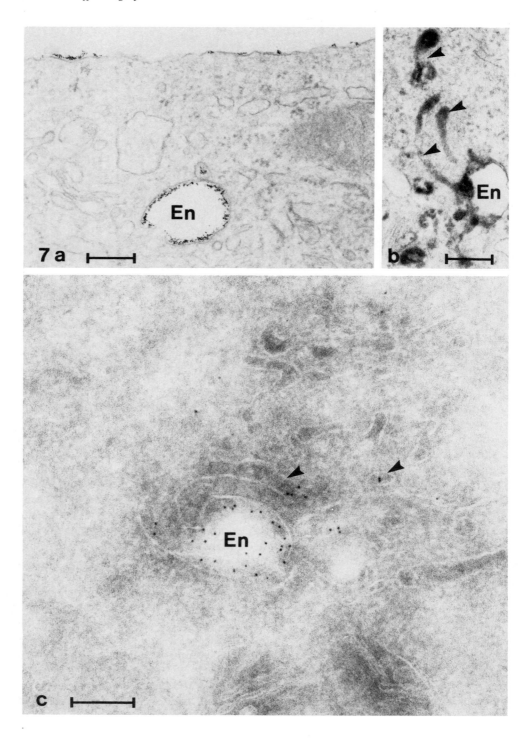

FIGURE 7.   Examples of endosomes containing internalized ricin: (a) shows a vacuolar endosome (En) of a Vero cell after 15 min of incubation with ricin-gold (5 nm) at 37°C; (b) shows a vacuolar endosome (En) with associated tubular and vesicular elements (arrowheads) of cultured human breast epithelial cells incubated with monovalent (i.e., ligand-marker 1:1) ricin-HRP conjugate for 60 min at 37°C; (c) an ultracryosection of a BHK-21 cell incubated with ricin for 60 min at 37°C. A vacuolar endosome (En) and associated tubulovesicular structures (arrowheads) are seen, many of them containing ricin as detected immunocytochemically (antiricin followed by 8-nm PAG). Bars, 0.25 μm.

that ricin-gold, which is polyvalent, and both polyvalent and monovalent ricin-HRP conjugates are all delivered to endosomes, and that only monovalent ricin-HRP—like unconjugated ricin—reaches the Golgi complex from the endosomal system (see Section VI). Polyvalent conjugates are delivered to lysosomes.[13] Similarly, Neutra et al.[94] found that the intracellular routing of transferrin is altered when this ligand is conjugated to gold. These observations are of obvious importance for the interpretation of ultrastructural studies where ligand conjugates (e.g., ligand-gold and ligand-HRP) with different "valency" are used.

Even though the internalization of surface-bound ricin is a slow process compared to, for example, the uptake of transferrin, ricin can be detected in peripheral endosomes after incubation for only a few minutes (see above). This indicates that ricin binding sites (terminal galactose) are present on a variety of cell surface molecules. It should be stressed that some "peripheral endosomes", appearing as round or elongate vacuolar profiles in random sections, may in fact not be endosomes, but invaginations of the cell surface (Figures 4 and 5) (see also Section IV).

Whereas specific markers for lysosomes (e.g., acid hydrolases, lysosomal membrane proteins) and the Golgi complex (e.g., galactosyl and sialyl transferases) exist, there are at present no specific markers for endosomes. It is possible, however, to classify as an endosome a structure that contains a labeled ligand, if the incubation is performed at 18 to 20°C rather than at 37°C.[14,15,75,95-100] At this low temperature, delivery of internalized molecules to lysosomes is strongly inhibited in a number of different cell types. However, a recent study showed that hydrolysis of endocytosed molecules begins within a few minutes at 37°C, and that exposure of endocytosed molecules to at least one lysosomal enzyme can take place below 20°C.[101] These data, therefore, suggest that the traditional distinction between an endosomal compartment, which is accessible to endocytosed molecules at low temperatures, and a lysosomal compartment, which is not, is too rigid. Rather, endosomes may gradually mature into lysosomes, and a temperature block could interrupt this maturation at a relatively late stage. The study by Roederer et al.[101] favors the concept of lysosomal maturation over the concept of a vesical shuttle between discrete endosomal and lysosomal compartments.[23] It may in general be more reasonable to use the term "endosome" as a collective for early endocytic compartments, and the term "lysosome" as a collective for late endocytic compartments on the line of maturation.

Biochemical data indicate that a proportion of internalized ricin molecules, perhaps those which detach from their binding sites in endosomes, are transferred to lysosomes. Here they are slowly degraded.[48] It is not clear whether this slow degradation is due to slow delivery to lysosomes or to the fact that ricin is very resistant to proteolytic enzymes. A number of compounds were found to inhibit the degradation of ricin by the cells (colchicine, cytochalasin B, and azide, as well as weak bases). None of these compounds inhibited the endocytosis of ricin or the recycling of the endocytosed toxin back to the cell surface.[48]

Also morphological studies with ricin conjugates and immunocytochemical detection of ricin have revealed the presence of this ligand in lysosome-like structures (Figure 8).[12,15] Using quantitative immunogold cytochemistry on ultracryosections, we found that after 60 min of ricin incubation, about 35% of the total amount of internalized ricin was present in lysosome-like structures[15] (Table 2). However, as will be discussed below (Section VII), we do not believe that the delivery of ricin to lysosomes is of any importance for the cytotoxic effect.

Considerable attention has been focused on the three-dimensional organization of the endosomal system in various cell types.[12,14,88,89,102-105] By using ricin-gold and monovalent ricin-HRP conjugates to trace endosomes combined with very thick (150 to 200 nm) or thin (20 to 50 nm) consecutive sections, we found that the endosomes are an extremely pleomorphic intracellular compartment (Figures 7 and 8).[12,14,138] Endosomes may appear as isolated vacuolar structures without any tubular appendices, or they may appear as a single

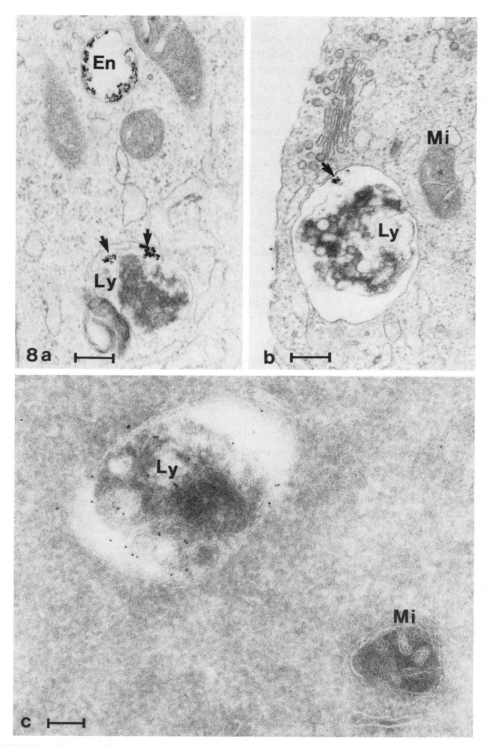

FIGURE 8.   Examples of lysosome-like structures containing internalized ricin: (a) and (b) show typical lysosomal structures (Ly) with ricin-gold (5 nm) in Vero cells after incubation with the conjugate for 60 min at 37°C. In (a) is also seen an endosome (En) containing ricin. (c) is an ultracryosection of a BHK-21 cell incubated for 60 min at 37°C with unconjugated ricin. Following immunogold labeling (8-nm PAG) ricin is seen in a lysosome (Ly). Note that whereas ricin-gold appears aggregated in the lysosomes in (a) and (b) (arrows), unconjugated ricin is scattered throughout the lysosome in (c). Mi, mitochondria; bars, 0.25 μm.

TABLE 2
Quantitation of Internalized Ricin in Various Compartments of BHK-21 Cells as Detected by Immunogold Cytochemistry[a]

Compartment	Amount of gold particles per compartment per cell[b]	% of total gold	No. of ricin molecules
Golgi stacks[c]	410	1.2	2.3×10^4
Trans-Golgi network[c]	1,012	2.9	5.5×10^4
Endosomes	22,057	63.0	1.2×10^6
Lysosomes	11,517	32.9	6.3×10^5
	Total ca. 3.5×10^4 gold particles per cell		Total ca. 1.9×10^6 molecules per cell[d]

[a] Data modified from Reference 15. Cells were infected with VSV ts 045, and the G protein of the virus was used as a marker of the biosynthetic pathway. The cells were also incubated with ricin for 1 h at 39.5°C followed by 2 h at 19.5°C in the presence of ricin (see text).

[b] These values were calculated using stereology (see Reference 15).

[c] Ricin in these two compartments was determined by immunocytochemical colocalization with G protein.

[d] This value was measured biochemically using ^{125}I-ricin (see Reference 15).

tubule or a group of isolated tubular structures. Mostly, however, the endosomal system was found to comprise groups of intermixed vacuoles, tubular or cisternal structures, and small vesicles (Figure 7). Very often, but not always, the tubular/cisternal structures were connected with the larger vacuolar elements. Accordingly, what may appear as isolated endosomal structures in a random section often turn out to be interconnected when analyzed using serial sectioning. It was also apparent that the individual groups of more or less interconnected endosomal vacuoles, tubules and vesicles represented discrete elements of the total endosomal apparatus of the cells. Using HRP as a fluid-phase marker of endocytosis in BHK cells and computer analysis of serial sections, Marsh et al.[105] found a similar three-dimensional organization of the endosomal system.

While the gross three-dimensional organization of the endosomal system is readily revealed by analyzing realtively thick sections, care must be taken to adjust the section thickness to determine the spatial relations between the smallest structures of interest. Thus, in order to decide whether a small vesicular structure close to an endosomal vacuole is actually connected with the vacuole, or localized freely in the cytoplasm, very thin sections (about 20 to 25 nm) must be used.[12] Some details of the application of serial section analysis, in particular with respect to the importance of the section thickness, are discussed elsewhere.[76,77,106-108]

VI. SOME INTERNALIZED RICIN MOLECULES ARE DELIVERED TO THE *TRANS*-GOLGI NETWORK

It is apparent from numerous EM studies that internalized molecules, such as HRP and cationized ferritin (CF), reach Golgi-associated compartments in some cell types[109-116] while this does not seem to be the case in other cell types.[80] Ricin-HRP conjugates were first shown in Golgi compartments in neuroblastoma cells,[58,59] and more recently in MCF-7, T47D, and Vero cells (Figures 9 and 10).[13,14,117] However, two major problems exist for such studies. First, without using specific markers for the secretory-Golgi pathway, it is not possible to determine unequivocally whether a Golgi-associated compartment labeled by an endocytosed, exogenous marker belongs to the secretory-Golgi pathway, or whether it represents an endosomal compartment.[14,15,118] It is well established that the endosomal system is highly pleomorphic and comprises vacuolar structures as well as tubulovesicular structures (see Section V), some of which can be found in close proximity to the Golgi complex.[14,105,119]

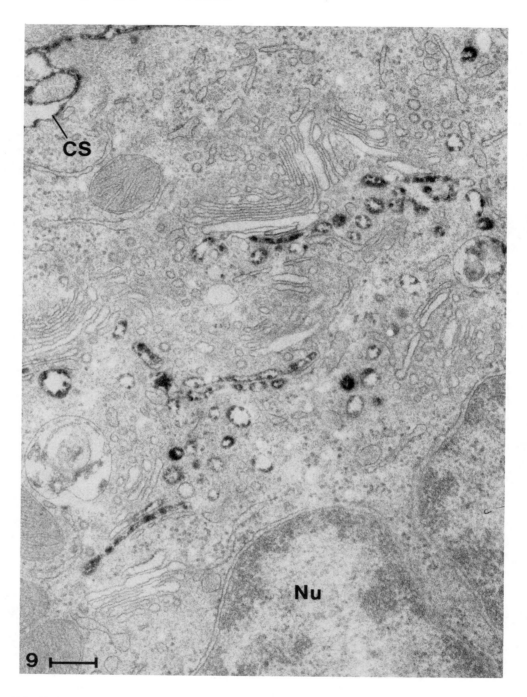

FIGURE 9.   Portion of an MCF-7 cell between the cell surface (CS) and the nucleus (Nu) showing extensive Golgi profiles. Cells were incubated with a monovalent ricin-HRP conjugate for 30 min at 37°C, and many Golgi-associated structures are distinctly labeled by the conjugate. Bar, 0.25 μm.

Second, even though a Golgi-associated compartment containing endocytosed molecules can be shown to belong to the secretory pathway, as was convincingly documented in a few studies of special cell types with a characteristic secretory product in Golgi cisternae as well as in budding and free secretory vesicles,[109,120] it is not certain what fraction of the total amount on internalized molecules reaches these secretory Golgi elements.

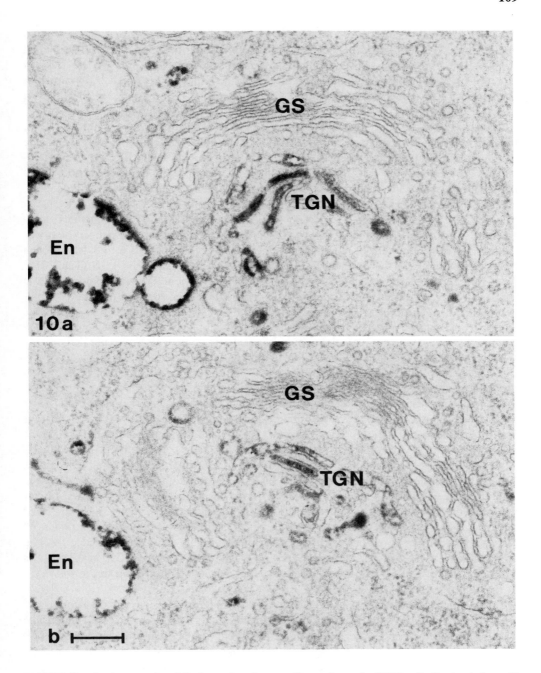

FIGURE 10. Sections no. 4 and 8 of a series of consecutive sections of a T47D cell after incubation with monovalent ricin-HRP for 60 min at 37°C. Ricin is seen in typical TGN profiles as well as in an endosome (En), while the Golgi stack (GS) is unlabeled. Bar, 0.25 μm.

To study this, we[15] recently used the glycosylated 57-kDa membrane-spanning G protein of vesicular stomatitis virus (VSV)[121] as a marker of the secretory-Golgi pathway of cells exposed to ricin. When cells are infected with VSV, they synthesize the G protein, which is subsequently transported within about half an hour from the endoplasmic reticulum (ER) via the Golgi complex to the cell surface.[121] In the case of the temperature sensitive mutant ts 045 of VSV which we used,[15] the G protein is synthesized at the nonpermissive temperature, 39.5°C, but remains in the ER at this temperature. On the other hand, at the permissive temperature (31°C), the G protein is rapidly transported via the Golgi complex to the cell surface. This transport can be stopped by a temperature block. Thus, at 19 to 20°C, the G protein leaves the ER and accumulates in large amounts in the Golgi complex, in particular in the trans-Golgi network (TGN),[119] and very little or no G protein reaches the cell surface.[15,99]

When cells infected with VSV ts 045 are incubated with ricin at 39.5°C for up to 1 h, internalized ligand is present in various endosomal-lysosomal structures as well as in some Golgi-associated structures, whereas G protein is only present in the ER. When the temperature is then lowered to 19 to 20°C, no more ricin will reach Golgi-associated structures,[14,117] but now G protein moves to the Golgi complex. By using quantitative double-labeling immunogold cytochemistry on ultrathin cryosections of VSV ts 045-infected BHK cells incubated with ricin for 60 min (Figure 11), in combination with biochemical measurements, we could show that about 4 to 5% of the total amount of internalized ricin was colocalized with G protein in the Golgi complex. This corresponds to 6 to 8 × 10^4 ricin molecules per cell. Of this amount, at least 75 to 80% was found in the TGN (Table 2).[15]

Using preembedding immunoperoxidase cytochemistry, ricin has been demonstrated in similar Golgi compartments.[13] However, when ricin conjugates are used instead of immunocytochemical detection of ricin, conflicting results have been obtained.[13] As mentioned above, we found that whereas endocytosed ricin-gold conjugates and polyvalent ricin-HRP conjugates (i.e., conjugates containing more than one ricin molecule) were not delivered to Golgi-associated compartments to any notable extent, monovalent ricin-HRP was readily found there.[12-14] These observations suggest that a sorting step has taken place in the endosome based on the valency of the ligand or ligand conjugate.

Another interesting aspect related to the organization of the Golgi complex, which was clearly revealed during the studies with monovalent ricin HRP mentioned above, is that *cis* and *trans* faces of the Golgi stack (as revealed in random, single sections) cannot be determined simply by relating them to the convex and concave sides of the stack, respectively. Although this oversimplified association between the functional polarity of the Golgi complex and its (apparent) curvature is often used in the literature, we found, by using serial section analysis, that one and the same ricin-labeled, Golgi-associated structure localized to, for instance, the concave side of the complex at one end of the series of sections and appeared on the convex side at the other end of the series.[13] The Golgi stacks show a pronounced "wavy" three-dimensional organization. Therefore, not until we used the double-labeling protocol with a *trans*-Golgi marker (G protein accumulated in the TGN at 19.5°C) as described above, could we localize unequivocally most internalized ricin to the trans-side of the Golgi.

Altogether, these studies show that (1) a sorting of internalized ricin and ricin conjugates with respect to further routing, for example, to the Golgi, takes place in the endosomal system. The sorting is at least in part based on the valency of the ricin conjugates. Therefore, in studies on transport of endocytosed toxin (and other ligand) conjugates, the selection among various conjugates with respect to valency (i.e., how many binding sites they have) is of crucial importance (see also Reference 94). (2) The cis and trans sides of the Golgi cannot be determined by the curvature of the stack, and (3) —most importantly—a proportion of the internalized native (i.e., unconjugated) ricin does reach secretory-Golgi compartments, in particular the TGN.

The observations that only a little ricin is detached from its binding sites at the low pH associated with endocytic compartments[15] (see Section V.), and that some internalized ricin reaches the TGN, indicate that plasma membrane molecules with terminal galactosyl residues (glycoproteins and glycolipids) are delivered to this last Golgi compartment where the sialyl transferases are probably localized.[119,122,123] This also may be true when such membrane

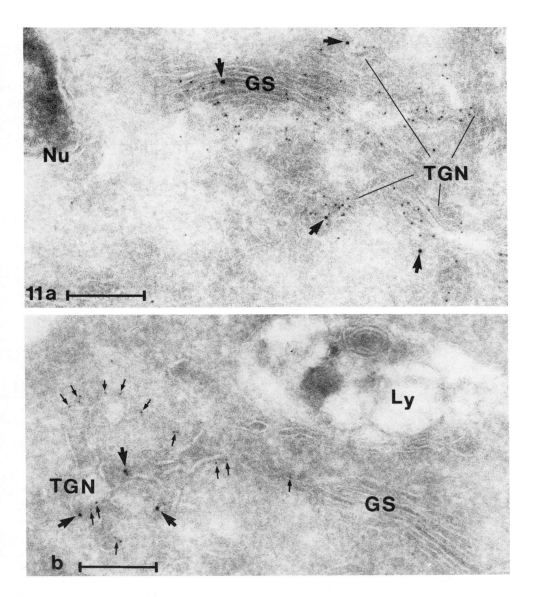

FIGURE 11.   Immunogold labeling of ultracryosections to detect internalized ricin (large, 10-nm golds) and VSV G-protein (small, 6-nm golds) simultaneously. BHK-21 cells were infected with the ts 045 VSV mutant and further incubated for 60 min at 39,5°C with ricin, as described in Section VI. Ricin (large arrows) co-localizes with G-protein (small arrows in [b]) — which is a marker of the biosynthetic-exocytic pathway under the present experimental conditions — mainly in the TGN. GS, Golgi stack; Ly, lysosome; Nu, nucleus; bars, 0.25 μm.

molecules are not tagged by ricin. In the TGN, internalized, recycling membrane molecules (such as receptors) may have the opportunity of being (re)sialylated *en route*. Most likely, only a small proportion of the total amount of internalized membrane molecules would at any given time point use the Golgi pathway for recycling, whereas the majority of the recycling molecules seems to be somehow selected for a rapid recycling directly from the endosomal system (see Section V). Probably the strongest support for this view has been presented by Snider and Rogers[124] who found that desialylated transferrin receptors, following internalization, slowly returned to the cell surface in a sialylated form. This shows that at least a fraction of internalized transferrin receptors (a receptor which is known to recycle perhaps several hundred times during its lifetime) are recycled via the TGN. In a more recent study, Snider and Rogers[125] provided evidence that some recycling plasma membrane molecules may not only reach the Golgi compartment(s) that contain(s) sialyl transferase, but also earlier Golgi compartments housing mannosidases.

VII. EVIDENCE THAT TRANSLOCATION OF RICIN A-CHAIN TAKES PLACE IN THE *TRANS*-GOLGI NETWORK

The mechanism by which internalized ricin is translocated to the cytosol where the A-chain inactivates the ribosomal 60S subunit is not clear. In a series of classical studies using ricin-ferritin conjugates, Nicolson and coworkers[38,126-128] found the conjugate in large endocytic vacuoles (presumably endosomes and lysosomes) as well as free in the cytosol. They, therefore, concluded that the mechanism by which ricin conjugates (and native ricin) reach the cytosol is by simple rupture of the membrane of endocytic vacuoles.

More recent studies have not been able to confirm this model. Rather, ricin A-chains, like the A fragment of diphtheria toxin and other bacterial and plant toxins, are probably somehow translocated across the membrane of one (or more) intact intracellular compartment(s). This compartment does not seem to be the same for all toxins. Thus, the A fragment of diphtheria toxin is translocated rapidly after uptake of the toxin, presumable in endosomes. The low pH in endosomes favors this process.[49,50,52,129,130] In contrast, it takes about 60 to 90 min after initial uptake of ricin before its cytotoxic effect can be measured biochemically as a reduction in the incorporation of ^{3}H-leucine (see, for instance Figure 4 in Reference 15). Therefore, it seems likely that the ricin A chain is translocated from a compartment localized distal to the endosomes in the endocytic pathway. Also, it is known that low pH as found in endosomes and lysosomes is not required for ricin A-chain translocation (see above, Section V). Although it cannot be ruled out that ricin A-chains are translocated from lysosomes, this does not appear likely. When the pH in lysosomes is increased by treatment of cells with 10 mM NH$_4$Cl, the cells are sensitized to ricin.[55] Therefore, a normal lysosomal function is not required for ricin entry into the cytosol.

We have recently shown that incubation of cells at 18°C rather than at 37°C has only a slight effect on the sensitivity of cells to diphtheria toxin, whereas the sensitivity to ricin is almost completely abolished under the same conditions.[14] Also, at 18°C, no ricin reaches the Golgi complex.[14,15] These data, therefore, point to a Golgi compartment, most likely the TGN, where the majority of internalized ricin molecules in the Golgi complex are found, as the site of ricin A-chain translocation. Treatment of cells with low concentrations of monensin (0.1 μM) sensitizes them to ricin,[48] although it has no measurable effect on the pH of endosomes and lysosomes.[131] However, such low concentrations of monensin induce morphological changes in the Golgi complex.[132]

As mentioned above, optimal ricin A-chain translocation requires a pH which is not very acidic. Although the TGN is thought to have a lower pH than the cytosol, as revealed with the DAMP technique,[133,134] the TGN is probably only mildly acidic. Thus, recent studies by Boulay et al.[135] showed that the fusion-incompetent precursor of the membrane-fusion protein, hemagglutinin (HA), of influenza virus is sensitive to low pH. Thus pH <6.0 to 6.3 induces irreversible conformational changes in the precursor. This may indicate that no compartment on the biosynthetic pathway, including the TGN, has a pH <6.0.

Moreover, cycloheximide, which blocks the peptidyl transferase reaction on ribosomes, sensitizes cells to ricin.[15,117] The reason for this could be that ricin under normal conditions competes with numerous newly synthetized molecules in the Golgi complex for binding sites and/or some enzymatic processing necessary for A-chain translocation to occur. Recently Youle and Colombatti[136] showed that a hybridoma cell secreting antibodies against ricin is

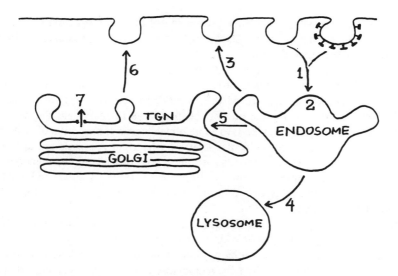

FIGURE 12. Intracellular routing and sorting of ricin. Ricin, bound to membrane glycoproteins and glycolipids, is internalized via both uncoated and coated pits and vesicles (1) to reach endosomes (2). From endosomes, ricin may be rapidly recycled (3), be delivered to lysosomes (4), or be delivered to the trans-Golgi network (TGN) (5) from where it may be recycled (6). Translocation of ricin A-chain is thought to take place in the TGN (7).

resistant to the toxin. A likely explanation of this observation is that the internalized ricin is inactivated intracellularly by the antiricin antibodies during their transport through the Golgi complex.[136] Only about 1% of the total amount of internalized ricin seems to be translocated to the cytosol.[136] Our finding that about 5% of the total amount of internalized ricin (in BHK cells exposed to ricin for 60 min) reaches the Golgi complex and mainly the TGN,[15] is in accordance with the hypothesis that this Golgi compartment is the site of ricin A-chain translocation.

In conclusion, even though other possibilities cannot be completely ruled out at the moment, much evidence has accumulated recently in favor of the concept that ricin must reach the TGN to exert its cytotoxic effect.

VIII. CONCLUSIONS AND PERSPECTIVES

A schematic presentation of our current knowledge and speculations about endocytosis and intracellular pathways of ricin leading to death of the cell is shown in Figure 12.

Binding sites or receptors (membrane glycoproteins and glycolipids) tagged by ricin are numerous all over the cell surface. Some of them are rapidly internalized while others may not be endocytosed at all. On average, the uptake of cell-bound ricin is a slow process, compared to that of, for instance, transferrin. Both coated and smooth pits and vesicles may participate in the uptake of ricin. Thereafter, ricin reaches the pleomorphic endosomal system where most ricin molecules remain attached to the membrane binding sites irrespective of the low pH prevailing here.

Some ricin molecules, presumably attached to the binding sites, are recycled to the cell surface by small vesicular and tubular structures budding off from the membrane of the endosomal system. Other ricin molecules, perhaps those which detach from the binding sites at the low pH, are transported to lysosomes where they are slowly degraded. More interestingly, some ricin molecules follow their membrane binding sites to the Golgi complex, in particular to the TGN, from where they may be routed back to the cell surface along the secretory pathway, possibly after some processing of the receptor. Finally, the A-chains of some ricin molecules that reach the TGN may be translocated across the membrane to reach the cytosol and there inhibit protein synthesis.

Many of these steps need further experimental elucidation, in particular in relation to the function and application of immunotoxins. Thus, the existence of an alternative (uncoated) endocytic pathway must be proven. Also, it would be interesting to know whether ricin has to be internalized specifically via one of the pathways to intoxicate the cell, or whether endocytosis along both pathways leads to an identical degree of inhibition of protein synthesis. A better understanding at the molecular level of why ricin A-chains are apparently translocated from the TGN is warranted. Also, studies should focus on which endocytic pathway various ricin-containing immunotoxins with different toxicities follow, and whether there is a correlation between the transport of an immunotoxin to the Golgi complex and the degree of toxicity of the conjugate.

There is increasing interest in designing new and more powerful immunotoxins with both higher specificity and higher toxicity for use in cancer therapy. As far as the toxicity is concerned, we think that this will be possible *only* with more detailed information on the various intracellular stations visited by endocytosed ricin, and most important, the compartment(s) from which the toxic ricin A-chain is translocated and to which the A-chain of a ricin immunotoxin should, therefore, be delivered.

Moreover, it is our hope that future studies, which combine morphological and biochemical techniques of the cellular handling of various toxins, will increase our understanding of the complex picture of endocytosis we have at the moment.

ACKNOWLEDGMENTS

The work from our laboratories referred to in this review has been supported by the NOVO Foundation, the Danish Medical Research Council, the European Molecular Biology Organization (EMBO), and the Norwegian Cancer Society.

REFERENCES

1. **Goldstein, J. L., Brown, M. S., Anderson, R. G. W., Russell, D. W., and Schneider, W. J.,** Receptor-mediated endocytosis: concepts emerging from the LDL receptor system, *Annu. Rev. Cell Biol.,* 1, 1, 1985.
2. **Wileman, T., Harding, C., and Stahl, P.,** Receptor-mediated endocytosis, *Biochem. J.,* 232, 1, 1985.
3. **Helenius, A., Kartenbeck, J., Simons, K., and Fries, E.,** On the entry of Semliki Forest Virus into BHK-21 cells, *J. Cell Biol.,* 84, 404, 1980.
4. **Madshus, I. H., Sandvig, K., Olsnes, S., and van Deurs, B.,** Effect of reduced endocytosis induced by hypotonic shock and potassium depletion on the infection of Hep-2 cells by Picornaviruses, *J. Cell. Physiol.,* 131, 14, 1987.
5. **Marsh, M.,** The entry of enveloped viruses into cells by endocytosis, *Biochem. J.,* 218, 1, 1984.
6. **Marsh, M. and Helenius, A.,** Adsorptive endocytosis of Semliki Forest Virus, *J. Mol. Biol.,* 142, 439, 1980.
7. **Olsnes, S. and Pihl, A.,** Toxic lectins and related proteins, in *Molecular Action of Toxins and Viruses,* Cohen, P. and van Heyningen, S., Eds., Elsevier/North Holland, Amsterdam, 1982, 51.
8. **Olsnes, S. and Sandvig, K.,** Entry of toxic proteins into cells, in *Receptor-Mediated Endocytosis: Receptors and Recognition,* Series B, Vol. 15, Cuatrecasas, P. and Roth, T. F., Eds., Chapman & Hall, London, 1983, 188.
9. **Olsnes, S. and Sandvig, K.,** Entry of polypeptide toxins into animal cells, in *Endocytosis,* Pastan, I. and Willingham, M. C., Eds., Plenum Press, New York, 1985, 195.
10. **Gonatas, N. K., Stieber, A., Kim, S. U., Graham, D. I., and Avrameas, S.,** Internalization of neuronal membrane ricin receptors into the Golgi apparatus, *Exp. Cell Res.,* 94, 426, 1975.

11. **Hickey, W. F., Stieber, A., Hogue-Angeletti, R., Gonatas, J., and Gonatas, N. K.**, Nerve growth factor induced changes in the Golgi apparatus of PC-12 rat pheochromocytoma cells as studied by ligand endocytosis, cytochemical and morphometric methods, *J. Neurocytol.*, 12, 751, 1983.

12. **van Deurs, B., Pedersen, L. R., Sundan, A., Olsnes, S., and Sandvig, K.**, Receptor-mediated endocytosis of a ricin-colloidal gold conjugate in Vero cells. Intracellular routing to vacuolar and tubulo-vesicular portions of the endosomal system, *Exp. Cell Res.*, 159, 287, 1985.

13. **van Deurs, B., Tønnessen, T. I., Petersen, O. W., Sandvig, K., and Olsnes, S.**, Routing of internalized ricin and ricin conjugates to the Golgi complex, *J. Cell Biol.*, 102, 37, 1986.

14. **van Deurs, B., Petersen, O. W., Olsnes, S., and Sandvig, K.**, Delivery of internalized ricin from endosomes to cisternal Golgi elements is a discontinuous, temperature sensitive process, *Exp. Cell Res.*, 171, 137, 1987.

15. **van Deurs, B., Sandvig, K., Petersen, O. W., Olsnes, S., Simons, K., and Griffiths, G.**, Estimation of the amount of internalized ricin that reaches the trans Golgi network, *J. Cell Biol.*, 106, 253, 1988.

16. **Sandvig, K., Olsnes, S., Petersen, O. W., and van Deurs, B.**, Acidification of the cytosol inhibits endocytosis from coated pits, *J. Cell Biol.*, 105, 679, 1987.

17. **Olsnes, S., Sandvig, K., Moskaug, J. Ø., Stenmark, H., and van Deurs, B.**, Toxin translocation across membranes and intracellular mechanisms of action. *Proc. Int. Symp. Workshop Verocytotoxin*, Toronto, 1987, in press.

18. **Vitetta, E. S., Cushley, W., and Uhr, J. W.**, Synergy of ricin A chain-containing immunotoxins and ricin B chain-containing immunotoxins in in vitro killing of neoplastic human B cells, *Proc. Natl. Acad. Sci. U.S.A.*, 80, 6332, 1983.

19. **Uhr, J. W.**, Immunotoxins: Harnessing natures poisons, *J. Immunol.*, 133, 1, 1984.

20. **Vitetta, E. S. and Uhr, J. W.**, Immunotoxins: redirecting natures poisons, *Cell*, 41, 653, 1985.

21. **Pastan, I., Willingham, M. C., and FitzGerald, D. J. P.**, Immunotoxins, *Cell*, 47, 641, 1986.

22. **Anderson, R. G. W. and Kaplan, J.**, Receptor-mediated endocytosis, *Mod. Cell Biol.*, 1, 1, 1983.

23. **Helenius, A., Mellman, I., Wall, D., and Hubbard, A.**, Endosomes, *Trends Biochem. Sci.*, 8, 245, 1983.

24. **Mellman, I.**, Membrane recycling during endocytosis, in *Lysosomes in Biology and Pathology*, Dingle, J. T., Dean, R. T., and Sly, W., Eds., Elsevier, Amsterdam, 1984, 201.

25. **Pastan, I. and Willingham, M. C.**, The pathway of endocytosis, in *Endocytosis*, Pastan, I., and Willingham, M. C., Eds., Plenum Press, New York, 1985, 1.

26. **Steinman, R. M., Mellman, I. S., Muller, W. A., and Cohn, Z. A.**, Endocytosis and recycling of plasma membrane, *J. Cell Biol.*, 96, 1, 1983.

27. **Willingham, M. C. and Pastan, I.**, Endocytosis and exocytosis: current concepts of vesicle traffic in animal cells, *Int. Rev. Cytol.*, 92, 51, 1984.

28. **Youle, R. J., and Huang, A. H. C.**, Protein bodies from the endosperm of castor beans. Subfractionation, protein components, lectins, and changes during germination, *Plant Physiol.*, 58, 703, 1976.

29. **Lord, J. M.**, Synthesis and intracellular transport of lectin and storage protein precursors in endosperm from castor bean, *Eur. J. Biochem.*, 146, 403, 1985.

30. **Harley, S. M. and Lord, J. M.**, *In vitro* endoproteolytic cleavage of castor bean lectin precursors, *Plant Sci.*, 41, 111, 1985.

31. **Zentz, C., Frenoy, J.-P., and Bourrillon, R.**, Binding of galactose and lactose to ricin. Equilibrium studies, *Biochim. Biophys. Acta*, 536, 18, 1978.

32. **Endo, Y., Mitsui, K., Motizuki, M., and Tsurugi, K.**, The mechanism of action of ricin and related toxic lectins on eukaryotic ribosomes. The site and the characteristics of the modification in 28 S ribosomal RNA caused by the toxins, *J. Biol. Chem.*, 262, 5908, 1987.

33. **Endo, Y. and Tsurugi, T.**, RNA N-glycosidase activity of ricin A-chain. Mechanism of action of the toxic lectin ricin on eucaryotic ribosomes, *J. Biol. Chem.*, 262, 8128, 1987.

34. **Calderwood, S. B., Auclair, F., Donohue-Rolfe, A., Keusch, G. T., and Mekalanos, J. J.**, Nucleotide sequence of the Shiga-like toxin genes of *Escherichia coli*, *Proc. Natl. Acad. Sci. U.S.A.*, 84, 4364, 1987.

35. **Jackson, M. P., Neill, R. J., O'Brian, A. D., Holmes, R. K., and Newland, J. W.**, Nucleotide sequence analysis of the structural gene for Shiga-like toxin I encoded by bacteriophage 933 J from *Escherichia coli* 933, *Microbiol. Pathogen.*, 2, 147, 1987.

36. **Sandvig, K., Olsnes, S., and Pihl, A.**, Kinetics of binding of the toxic lectins abrin and ricin to surface receptors on human cells, *J. Biol. Chem.*, 251, 3997, 1976.

37. **Sandvig, K., Olsnes, S., and Pihl, A.**, Binding, uptake and degradation of the toxic proteins abrin and ricin by toxin-resistant cell variants, *Eur. J. Biochem.*, 82, 13, 1978.

38. **Nicolson, G. L., Lacorbiere, M., and Eckhart, W.**, Qualitative and quantitative interactions of lectins with untreated and neuraminidase treated normal, wild-type, and temperature sensitive polyoma transformed fibroblasts, *Biochemistry*, 14, 172, 1975.

39. **Hughes, R. C. and Gardas, A.**, Phenotypic reversion of ricin-resistant hamster fibroblasts to a sensitive state after coating with glycolipid receptors, *Nature*, 264, 63, 1976.

40. **Sandvig, K., Olsnes, S., and Pihl, A.,** Chemical modifications of the toxic lectins abrin and ricin, *Eur. J. Biochem.*, 84, 323, 1978.

41. **Montford, W., Villafranca, J. E., Monzingo, A. F., Ernst, S. R., Katzin, B., Rutenber, E., Xuong, N. H., Hamlin, R., and Robertus, J. D.,** The three-dimensional structure of ricin at 2.8 Å, *J. Biol. Chem.*, 262, 5398, 1987.

42. **Simmonds, B. M., Stahl, P. D., and Russel, J. H.,** Mannose receptor-chain by macrophages. Multiple intracellular pathways for A chain translocation, *J. Biol. Chem.*, 261, 7912, 1986.

43. **Skilleter, D. N. and Foxwell, B. M. J.,** Selective uptake of ricin A-chain by hepatic non-parenchymal cells *in vitro*. Importance of mannose oligosaccharides in the toxin, *FEBS Lett.*, 192, 344, 1986.

44. **Foxwell, B. M. J., Blakey, D. C., Brown, A. N. F., Donovan, T. A. and Thorpe, P. E.,** The preparation of deglycosylated ricin by recombination of glycosidase-treated A- and B-chains: effects of deglycosylation on toxicity and *in vivo* distribution, *Biochim. Biophys. Acta*, 923, 59, 1987.

45. **Carlsson, J., Drevin, H., and Axen, R.,** Protein thiolation and reversible protein-protein conjugation. N-succinimidyl 3-(2- pyridyldithio)propionate, a new heterobifunctional reagent, *Biochem. J.*, 173, 723, 1978.

46. **Thorpe, P. E. and Ross, W. C. J.,** The preparation and cytotoxic properties of antibody-toxin conjugates, *Immunol., Rev.*, 62, 119, 1982.

47. **Dunn, W. A. and Hubbard, A. L.,** Receptor-mediated endocytosis of epidermal growth factor by hepatocytes in the perfused rat liver: ligand and receptor dynamics, *J. Cell Biol.*, 98, 2148, 1984.

48. **Sandvig, K. and Olsnes, S.,** Effect of temperature on the uptake, excretion and degradation of abrin and ricin by HeLa cells, *Exp. Cell Res.*, 121, 15, 1979.

49. **Draper, R. and Simon, M.,** The entry of diphtheria toxin into the mammalian cell cytoplasm: Evidence for lysosomal involvement, *J. Cell Biol.*, 87, 849, 1980.

50. **Sandvig, K. and Olsnes, S.,** Diphtheria toxin entry into cells is facilitated by low pH, *J. Cell Biol.*, 87, 828, 1980.

51. **Mellman, I., Fuchs, R., and Helenius, A.,** Acidification of the endocytic and exocytic pathways, *Annu. Rev. Biochem.*, 55, 663, 1986.

52. **Sandvig, K. and Olsnes, S.,** Rapid entry of nicked diphtheria toxin into cells at low pH. Characterization of the entry process and effects of low pH on the toxin molecule, *J. Biol. Chem.*, 256, 9068, 1981.

53. **Sundan, A., Sandvig, K., and Olsnes, S.,** Calmodulin antagonists sensitize cells to pseudomonas toxin, *J. Cell Physiol.*, 119, 15, 1984.

54. **Saelinger, C. B., Morris, R. E., and Foertsch, G.,** Trafficking of Pseudomonas Exotoxin A in mammalian cells, *Eur. J. Clin. Microbiol.*, 4, 170, 1985.

55. **Sandvig, K., Olsnes, S., and Pihl, A.,** Inhibitory effect of ammonium chloride and chloroquine on the entry of the toxic lectin modeccin into HeLa cells, *Biochem. Biophys. Res. Commun.*, 90, 648, 1979.

56. **Sandvig, K. and Olsnes, S.,** Entry of the toxin proteins abrin, modeccin, ricin and diphtheria toxin into cells. II. Effect of pH, metabolic inhibitors and ionophores and evidence for penetration from endocytic vesicles, *J. Biol. Chem.*, 257, 7504, 1982.

57. **Sandvig, K. and Olsnes, S.,** Entry of the toxic proteins abrin, ricin, modeccin and diphtheria toxin into cells. I. Requirement for calcium, *J. Biol. Chem.*, 257, 7495, 1982.

58. **Gonatas, N. K., Kim, S. U., Stieber, A., and Avrameas, S.,** Internalization of lectins in neuronal GERL, *J. Cell Biol.*, 73, 1, 1977.

59. **Gonatas, J., Stieber, A., Olsnes, S., and Gonatas, N. K.,** Pathways involved in fluid phase and adsorptive endocytosis in neuroblastoma, *J. Cell Biol.*, 87, 579, 1980.

60. **Pearse, B. M. F. and Bretscher, M. S.,** Membrane recycling by coated vesicles, *Annu. Rev. Biochem.*, 50, 85, 1981.

61. **Salisbury, J. L., Condeelis, J. S., and Satir, P.,** Receptor-mediated endocytosis: Machinery and regulation of the clathrin-coated vesicle pathway, *Int. Rev. Exp. Pathol.*, 24, 1, 1983.

62. **Huet, C., Ash, J. F., and Singer, S. J.,** The antibody-induced clustering and endocytosis of HLA antigens on cultured human fibroblasts, *Cell*, 21, 429, 1980.

63. **Montesano, R., Roth, J., Robert, A., and Orci, L.,** Non-coated membrane invaginations are involved in binding and internalization of cholera and tetanus toxins, *Nature*, 296, 651, 1982.

64. **Hopkins, C. R., Miller, K., and Beardmore, J. M.,** Receptor-mediated endocytosis of transferrin and epidermal growth factor receptors: A comparison of constitutive and ligand-induced uptake, *J. Cell Sci. Suppl.*, 3, 173, 1985.

65. **Ghitescu, L., Fixman, A., Simionescu, M., and Simionescu, N.,** Specific binding sites for albumin restricted to plasmalemmal vesicles of continous capillary endothelium: Receptor-mediated transcytosis, *J. Cell Biol.*, 102, 1304, 1986.

66. **Larkin, J. M., Brown, M. S., Goldstein, J. L., and Anderson, R. G. W.,** Depletion of intracellular potassium arrests coated pit formation and receptor-mediated endocytosis in fibroblasts, *Cell*, 33, 273, 1983.

67. **Larkin, J. M., Donzell, W. C., and Anderson, R. G. W.,** Modulation of intracellular potassium and ATP: Effects on coated pit function in fibroblasts and hepatocytes, *J. Cell. Physiol.*, 124, 372, 1985.

68. **Larkin, J. M., Donzell, W. C., and Anderson, R. G. W.,** Potassium-dependent assembly of coated pits: New coated pits form as planar clathrin lattices, *J. Cell Biol.,* 103, 2619, 1986.

69. **Moya, M., Dautry-Varsat, A., Goud, B., Louvard, D., and Boquet, P.,** Inhibition of coated pit formation in Hep-2 cells blocks the cytotoxicity of diphtheria toxin but not that of ricin toxin, *J. Cell Biol.,* 101, 548, 1985.

70. **Ghosh, P. C., Wellner, R. B., Cragoe, E. J., and Wu, H. C.,** Enhancement of ricin cytotoxicity in chinese hamster ovary cells by depletion of intracellular K^+: Evidence for an Na^+/H^+ exchange system in chinese hamster ovary cells, *J. Cell Biol.,* 101, 350, 1985.

71. **Li, W., Ryser, H. J.-P., Mandel, R., and Shen, W.-C.,** A defect in cell-cell fusion is associated with a defective fluid-phase and normal receptor-mediated endocytosis in a mutant of LM fibroblasts, *J. Cell Biol.,* 101 (No. 5, part 2), 421a, 1985.

72. **Li, W., Ryser, H. J.-P., and Shen, W.-C.,** Altered endocytosis in a mutant of LM fibroblasts defective in cell-cell fusion, *J. Cell. Physiol.,* 126, 161, 1986.

73. **Doxsey, S. J., Brodsky, F. M., Blank, G. S., and Helenius, A.,** Inhibition of endocytosis by anti-clathrin antibodies, *Cell,* 50, 453, 1987.

74. **Hopkins, C. R.,** Intracellular routing of transferrin and transferrin receptors in epidermoid carcinoma A431 cells, *Cell,* 35, 321, 1983.

75. **Hopkins, C. R. and Trowbridge, I. S.,** Internalization and processing of transferrin and the transferrin receptor in human carcinoma A431 cells, *J. Cell Biol.,* 97, 508, 1983.

76. **Bundgaard, M., Hagman, P., and Crone, C.,** The three-dimensional organization of plasmalemmal vesicular profiles in the endothelium of rat heart capillaries, *Microvasc. Res.,* 25, 358, 1983.

77. **Frøkjær-Jensen, J.,** Three-dimensional organization of plasmalemmal vesicles in endothelial cells. An analysis by serial sectioning of frog mesenteric capillaries, *J. Ultrastruct. Res.,* 73, 9, 1980.

78. **Noguchi, Y., Shibata, Y., and Yamamoto, T.,** Endothelial vesicular system in rapid-frozen muscle capillaries revealed by serial sectioning and deep etching, *Anat. Rec.,* 217, 355, 1987.

79. **Christensen, E. I.,** Rapid membrane recycling in renal proximal tubule cells, *Eur. J. Cell Biol.,* 29, 43, 1982.

80. **van Deurs, B. and Nilausen, K.,** Pinocytosis in mouse L-fibroblasts: Ultrastructural evidence for a direct membrane shuttle between the plasma membrane and the lysosomal compartment, *J. Cell Biol.,* 94, 279, 1982.

81. **van Deurs, B. and Christensen, E. I.,** Endocytosis in kidney proximal tubule cells and cultured fibroblasts: a review of the structural aspects of membrane recycling between the plasma membrane and endocytic vacuoles, *Eur. J. Cell Biol.,* 33, 163, 1984.

82. **Jarett, L. and Smith, R. M.,** Ultrastructural localization of insulin receptors in adipocytes, *Proc. Natl. Acad. Sci. U.S.A.,* 72, 3526, 1975.

83. **Carpentier, J.-L., Perrelet, A., and Orci, L.,** Morphological changes of the adipose cell plasma membrane during lipolysis, *J. Cell Biol.,* 72, 104, 1977.

84. **Brown, D., Weyer, P., and Orci, L.,** Nonclathrin-coated vesicles are involved in endocytosis in kidney collecting duct intercalated cells, *Anat. Rec.,* 218, 237, 1987.

85. **Brown, D. and Orci, L.,** The ''coat'' of kidney intercalated cell tubulovesicles does not contain clathrin, *Am. J. Physiol.,* 250, C605, 1986.

86. **Ciechanover, A., Schwartz, A. L., Dautry-Varsat, A., and Lodish, H. F.,** Kinetics of internalization and recycling of transferrin and the transferrin receptor in a human hepatoma cell line. Effect of lysosomotropic agents, *J. Biol. Chem.,* 258, 9681, 1983.

87. **Bleil, J. D. and Bretscher, M. S.,** Transferrin receptor and its recycling in HeLa cells, *EMBO J.,* 1, 351, 1982.

88. **Geuze, H. J., Slot, J. W., and Strous, G. J. A. M.,** Intracellular site of asialoglycoprotein receptor-ligand uncoupling: double-label immunoelectron microscopy during receptor-mediated endocytosis, *Cell,* 32, 277, 1983.

89. **Geuze, H. J., Slot, J. W., Strous, G. J. A. M., Peppard, J., von Figura, K., Hasilik, A., and Schwartz, A. L.,** Intracellular receptor sorting during endocytosis: Comparative immunoelectron microscopy of multiple receptors in rat liver, *Cell,* 37, 195, 1984.

90. **Anderson, R. G. W., Brown, M. S., Beisiegel, U., and Goldstein, J. L.,** Surface distribution and recycling of the low density lipoprotein receptor as visualized with antireceptor antibodies, *J. Cell Biol.,* 93, 523, 1982.

91. **Mellman, I., Plutner, H., and Ukkonen, P.,** Internalization and rapid recycling of macrophage Fc receptors tagged with monovalent antireceptor antibody: possible role of a prelysosomal compartment, *J. Cell Biol.,* 98, 1163, 1984.

92. **Mellman, I. and Plutner, H.,** Internalization and degradation of macrophage Fc receptors bound to polyvalent immune complexes, *J. Cell Biol.,* 98, 1170, 1984.

93. **Ukkonen, P., Lewis, V., Marsh, M., Helenius, A., and Mellman, I.,** Transport of macrophage Fc receptors and Fc receptor-bound ligands to lysosomes, *J. Exp. Med.,* 163, 952, 1986.

94. **Neutra, M. R., Ciechanover, A., Owen, L. S., and Lodish, H. F.,** Intracellular transport of transferrin-and asialoorosomucoid-colloidal gold conjugates to lysosomes after receptor-mediated endocytosis, *J. Histochem. Cytochem.,* 33, 1134, 1985.

95. **Dunn, W. A., Hubbard, A. L., and Aronson, N. A.,** Low temperature selectively inhibits fusion between pinocytic vesicles and lysosomes during heterophagy of ^{125}I-asialofetuin by the perfused rat liver, *J. Biol. Chem.,* 255, 5971, 1980.

96. **Dunn, W. A., Connolly, T. P., and Hubbard, A. L.,** Receptor-mediated endocytosis of epidermal growth factor by rat hepatocytes: receptor pathway, *J. Cell Biol.,* 102, 24, 1986.

97. **Marsh, M., Bolzau, E., and Helenius, A.,** Penetration of Semliki Forest Virus from acidic perlysosomal vacuoles, *Cell,* 32, 931, 1983.

98. **Miller, K., Beardmore, J., Kanety, H., Schlessinger, J., and Hopkins, C. R.,** Localization of the epidermal growth factor (EGF) receptor within the endosome of EGF-stimulated epidermoid carcinoma (A431) cells, *J. Cell Biol.,* 102, 500, 1986.

99. **Griffiths, G., Pfeiffer, S., Simons, K., and Matlin, K.,** Exit of newly synthesized membrane proteins from the trans cisterna of the Golgi complex to the plasma membrane, *J. Cell Biol.,* 101, 949, 1985.

100. **Sandvig, K., Sundan, A., and Olsnes, S.,** Evidence that modeccin and diphtheria toxin enter the cytosol from different vesicular compartments, *J. Cell Biol.,* 98, 963, 1984.

101. **Roederer, M., Bowser, R., and Murphy, R. F.,** Kinetics and temperature dependence of exposure of endocytosed material to proteolytic enzymes and low pH: evidence for a maturation model for the formation of lysosomes, *J. Cell. Physiol.,* 131, 200, 1987.

102. **Bretscher, M. S. and Thomson, J. N.,** The morphology of endosomes in giant HeLa cells, *Eur. J. Cell Biol.,* 37, 78, 1985.

103. **Hopkins, C. R.,** Membrane boundaries involved in the uptake and intracellular processing of cell surface receptors, *Trends Biochem. Sci.,* 11, 473, 1986.

104. **Hoppe, C. A., Connolly, T. P., and Hubbard, A. L.,** Transcellular transport of polymeric IgA in the rat hepatocyte: Biochemical and morphological characterization of the transport pathway, *J. Cell Biol.,* 101, 2113, 1985.

105. **Marsh, M., Griffiths, G., Dean, G. E., Mellman, I., and Helenius, A.,** Three-demensional structure of endosomes in BHK-21 cells, *Proc. Natl. Acad. Sci. U.S.A.,* 83, 2899, 1986.

106. **Petersen, O. W. and van Deurs, B.,** Serial-section analysis of coated pits and vesicles involved in adsorptive pinocytosis in cultured fibroblasts, *J. Cell Biol.,* 96, 277, 1983.

107. **van Deurs, B., Petersen, O. W., and Bundgaard, M.,** Do coated pinocytic vesicles exist?, *Trends Biochem. Sci.,* 8, 400, 1983.

108. **van Deurs, B., Petersen, O. W., and Bundgaard, M.,** Identification of free coated pinocytic vesicles in Swiss 3T3 cells, *EMBO J.* 3, 1959, 1984.

109. **Farquhar, M. G.,** Recovery of surface membrane in anterior pituitary cells. Variations in traffic detected with anionic and cationic ferritin, *J. Cell Biol.,* 78, R35, 1978.

110. **Farquhar, M. G. and Palade, G. E.,** The Golgi apparatus (complex) — (1954—1981) — from artifact to center stage, *J. Cell Biol.,* 91(3,Part 2), 77s, 1981.

111. **Herzog, V. and Miller, F.,** Membrane retrieval in epithelial cells of isolated thyroid follicles, *Eur. J. Cell Biol.,* 19, 203, 1979.

112. **Mata, L. R.,** Dynamics of HRPase absorption in the epithelial cells of the hamster seminal vesicles, *J. Microsc. Biol. Cell.,* 25, 127, 1976.

113. **Mata, L. R. and David-Ferreira, J. F.,** Transport of exogenous peroxidase to Golgi cisternae in the hamster seminal vesicle, *J. Microsc.,* 17, 103, 1973.

114. **Ottosen, P. D., Courtoy, P. J., and Farquhar, M. G.,** The pathways followed by membrane recovered from the surface of plasma cells and myeloma cells, *J. Exp. Med.,* 152, 1, 1980.

115. **Thyberg, J.,** Internalization of cationized ferritin into the golgi complex of cultured mouse peritoneal macrophages. Effects of colchicine and cytochalasin B, *Eur. J. Cell Biol.,* 23, 95, 1980.

116. **van Deurs, B., von Bulow, F., and Møller, M.,** Vesicular transport of cationized ferritin by the epithelium of the rat choroid plexus, *J. Cell Biol.,* 89, 131, 1981.

117. **Sandvig, K., Tønnessen, T. I., and Olsnes, S.,** Ability of inhibitors of glycosylation and protein synthesis to sensitize cells to abrin, ricin, Shigella toxin, and Pseudomonas toxin, *Cancer Res.,* 46, 6418, 1986.

118. **Orci, L., Ravazzola, M., Amherdt, M., Brown, D., and Perrelet, A.,** Transport of horseradish peroxidase from the cell surface to the Golgi in insulin-secreting cells: preferential labelling of cisternae located in an intermediate position of the stack, *EMBO J.,* 5, 2097, 1986.

119. **Griffiths, G. and Simons, K.,** The *trans* Golgi network: Sorting at the exit site of the Golgi complex, *Science,* 234, 438, 1986.

120. **Balin, B. J. and Broadwell, R. D.,** Lectin-labeled membrane is transferred to the Golgi complex in mouse pituitary cells *in vivo,* *J. Histochem. Cytochem.,* 35, 489, 1987.

121. **Dubois-Dalcq, M., Holmes, K. V., and Rentier, B.,** Assembly of enveloped RNA viruses, Springer-Verlag, Vienna, 1984.

122. **Fuller, S. D., Bravo, R., and Simons, K.,** An enzymatic assay reveals that proteins destined for the apical or basolateral domains of an epithelial cell line share the same late Golgi compartments, *EMBO J.,* 4, 297, 1985.

123. **Roth, J., Taatjes, D. J., Lucocq, J. M., Weinstein, J., and Paulson, J. C.,** Demonstration of an extensive trans-tubular network continuous with the Golgi apparatus stack that may function in glycosylation, *Cell,* 43, 287, 1985.

124. **Snider, M. D. and Rogers, O. C.,** Intracellular movement of cell surface receptors after endocytosis: resialylation of asialo-transferrin receptor in human erythroleukemia cells, *J. Cell Biol.,* 100, 826, 1985.

125. **Snider, M. D. and Rogers, O. C.,** Membrane traffic in animal cells: cellular glycoproteins return to the site of Golgi mannosidase I, *J. Cell Biol.,* 103, 265, 1986.

126. **Nicolson, G. L.,** Ultrastructural analysis of toxin binding and entry into mammalian cells, *Nature,* 251, 628, 1974.

127. **Nicolson, G. L., Lacorbiere, M., and Hunter, T. R.,** Mechanism of cell entry and toxicity of an affinity-purified lectin from Riccinus communis and its differential effects on normal and virus-transformed fibroblasts, *Cancer Res.,* 35, 144, 1975.

128. **Nicolson, G. L., Smith, J. R., and Hyman, R.,** Dynamics of toxin and lectin receptors on a hybridoma cell line and its toxin-resistant variant using ferritin-conjugated, ^{125}I-labeled ligand, *J. Cell Biol.,* 78, 565, 1978.

129. **Moskaug, J. Ø., Sandvig, K., and Olsnes, S.,** Cell-mediated reduction of the inter-fragment disulfide bond in nicked diphtheria toxin. A new method to study toxin entry to the cytosol, *J. Biol. Chem.,* 262, 10339, 1987.

130. **Moskaug, J. Ø., Sandvig, K., and Olsnes, S.,** Low pH-induced release of diphtheria toxin A-fragment to the cytosol. Biochemical evidence for transfer to the cytosol, *J. Biol. Chem.,* 263, 2518, 1988.

131. **Marsh, M., Wellsteed, J., Kern, H., Harms, E., and Helenius, A.,** Monensin inhibits Semliki Forest Virus penetration into culture cells, *Proc. Natl. Acad. Sci. U.S.A.,* 79, 5297, 1982.

132. **Tartakoff, A. and Vassalli, P.,** Plasma cell immunoglobulin M molecules. Their biosynthesis, assembly, and intracellular transport, *J. Cell Biol.,* 83, 284, 1979.

133. **Anderson, R. G. W. and Pathak, R. K.,** Vesicles and cisternae in the trans Golgi apparatus of human fibroblasts are acidic compartments, *Cell,* 40, 635, 1985.

134. **Schwartz, A. L., Strous, G. J. A. M., Slot, J. W., and Geuze, H. J.,** Immunoelectron microscopic localization of acidic intracellular compartments in hepatoma cells, *EMBO J.,* 4, 899, 1985.

135. **Boulay, F., Doms, R. W., Wilson, I., and Helenius, A.,** The influenza hemagglutinin precursor as an acid-sensitive probe of the biosynthetic pathway, *EMBO J.,* 6, 2643, 1987.

136. **Youle, R. J. and Colombatti, M.,** Hybridoma cells containing intracellular anti-ricin antibodies show ricin meets secretory antibody before entering the cytosol, *J. Biol. Chem.,* 262, 4676, 1987.

137. **Sandvig, K., Olsnes, S., Brown, J. E., Petersen, O. W., and van Deurs, B.,** Endocytosis from coated pits of Shiga toxin, a glycolipid-binding protein from *Shigella dysenteriae* 1, *J. Cell. Biol.,* 108, 1331, 1989.

Chapter 7

THE ANTHRAX TOXINS

Arthur M. Friedlander

TABLE OF CONTENTS

I. INTRODUCTION

Bacillus anthracis is the causative agent of anthrax. Anthrax is a serious, potentially fatal infection, which is usually confined to herbivores. Man may be infected via contact with diseased animals or animal products. To be considered virulent, the organism must produce both a poly-D-glutamic acid capsule and a three-component toxin; the components are protective antigen (PA), edema factor (EF), and lethal factor (LF). Once protective methods to control the disease were successful, research on anthrax slowed and has only been resumed in the last several years.

II. TOXIN STRUCTURE AND FUNCTION

B. anthracis produces two protein exotoxins: edema toxin and lethal toxin, which are important virulence factors in infection.[1,2] Both edema toxin and lethal toxin follow the general model for many protein toxins in possessing a binding or B component responsible for binding to receptors on target cells and an active or A component responsible for the toxin's biochemical activity.[3,4] However, the anthrax toxins are unusual in two respects. First, the B and A domains are located on separate noncovalently linked proteins, similar to the structure of a few other toxins, such as staphylococcal leukocidin[5] and botulinum C_2 toxin.[6] Second, the edema and lethal toxins share the identical B protein, a situation which is unique to anthrax. The B protein, called PA, has a mass of 83 kDa. This protein, when given together with a second protein, EF (89 kDa), comprises the edema toxin. The same B protein, protective antigen, in conjunction with a third protein, LF (83 kDa) constitutes the lethal toxin. Evidence gathered in the last several years has established that the three protein components of the two toxins are coded by three noncontiguous genes carried on a single 170 kilobase plasmid.[7-11] To date, the PA and EF genes have been sequenced and analysis shows no significant homology between them.[12,13]

The individual toxin components, PA, EF, or LF, by themselves show no biological activity in experimental animals. The edema toxin, PA + EF together, causes edema when injected into the skin of experimental animals[14,15] and is most likely responsible for the massive edema occurring in anthrax infections. The lethal toxin, PA + LF, is lethal for several animal species.[14,15] The idea that PA serves as the binding or B component for the lethal toxin had been suggested in early *in vivo* experiments[16] and LF had been shown to block the effect of EF given with PA.[14] The critical role of PA in resistance to infection was originally based upon its identification as an immunogen capable of protecting experimental animals against infection,[1,17] and by the fact that antiserum protects animals against the lethal toxin.[1] PA remains the basic component of the current human vaccines.[18] While the toxins are clearly critical virulence factors, their overall role in pathogenesis and immunity to infection remains to be definitively established.

The nature of the enzymatic activity of EF, the A protein component of edema toxin, was identified by Leppla as a calcium and calmodulin-dependent adenylate cyclase.[4,19] He observed that EF given together with PA caused massive elevations (>200 times basal levels) of cyclic AMP in eukaryotic cells in culture; this undoubtedly is responsible for the edema observed in clinical infections and for the abnormalities of phagocytic function that have been described.[20,21] There is no elevation of cyclic AMP when EF is given to intact cells in the absence of PA. Furthermore, excess LF was found to block the activity of EF added to cells in the presence of PA. These results led to the proposal for the current model of the anthrax toxins outlined above, confirming and extending the early *in vivo* observations.

Lethal toxin, PA + LF, was recently found to be cytolytic for mouse peritoneal macrophages, providing the necessary *in vitro* model to study the toxin's interaction with cells and mechanism of action.[22] The putative enzymatic activity of the A component of lethal toxin, the LF protein, has not yet been discovered. As with the edema toxin, both B and A proteins, PA + LF, are required for expression of biological activity on intact cells. LF is not toxic for cells in the absence of PA. While edema toxin is active and increases cyclic AMP levels in many cell types,[4] including macrophages, the cytolytic activity of lethal toxin is only observed in macrophages[22] or some macrophage-like cell lines.[23] Inhibition of growth of some nonmacrophage cell lines was observed when cells were plated at very low density[24] and some inhibition of protein synthesis can be measured after extended periods of time in certain other cell lines.[25]

III. INTERACTION OF TOXIN COMPONENTS WITH THE CELL SURFACE

Several lines of evidence suggest that PA binds to receptors on target cells, and then permits subsequent binding of LF or EF. This includes the fact that EF is able to inhibit lethal toxin activity in an animal lethality model.[26] Also, two monoclonal antibodies to PA have been prepared which neutralize both edema and lethal toxin activity.[27] These antibodies are specific for PA and did not bind to EF or LF. They inhibited PA binding to receptor sites on fetal Rhesus lung cells. Thus, work with monoclonal antibodies also provides circumstantial evidence that binding of PA to cell receptors is required for toxin activity.

In our initial studies using radioiodinated proteins, we directly measured the interaction of toxin components with target cells. We confirmed the basic B + A model for the anthrax toxins which had been based on biochemical or biological assays. The following overall model has emerged from these studies.[28,29] PA, the B protein, binds to cells by a single class of high affinity receptors (dissociation constant approximately $10^{-9} M/l$, 8,000 to 30,000 receptors per cell). The PA is then cleaved by a cell surface protease creating a binding site on the surface-bound PA to which LF or EF binds with high affinity. There is no binding of LF in the absence of PA; LF and EF appear to compete for the same site.

IV. INTRACELLULAR TRAFFICKING OF TOXIN

Knowledge concerning the internalization and intracellular trafficking of the anthrax toxins is just beginning to emerge. The observations about the interaction of toxin components with cell-surface receptors outlined above suggest that the overall internalization mechanism and trafficking for both edema toxin and lethal toxin is the same. PA appears to function as the common carrier for binding and transport of EF or LF to the cytosol.

The information on trafficking which is available is mainly derived from experiments with the lethal toxin on macrophages. Initial experiments demonstrated that lethal toxin is cytolytic with the earliest evidence of cell death occurring after 1-h exposure at 37°C.

Evidence from many investigators has established that some toxins, including diphtheria, pseudomonas exotoxin A, and modeccin,[30,31] as well as viruses[32,33] and protein ligands,[34] must pass through an acidic intracellular vesicular compartment before translocation to the cytosol where their biological activity is expressed. These conclusions showing the importance of low pH in translocation are based upon several experimental approaches which include: (1) manipulation of vesicular and cytosolic transmembrane pH; (2) low temperature which impairs the movement of ligands intracellularly; and (3) cell mutants defective in vesicular acidification. Using similar approaches, we have observed that lethal toxin must first be exposed to an acidic intracellular environment to express its cytotoxic effect on macrophages.[22] This study showed complete protection of macrophages from killing by lethal toxin when cells were pretreated with the amines, NH_4Cl and chloroquine, or with the proton ionophore monensin, which rapidly dissipate proton gradients and alkalinize intracellular acidic vesicles.[35,36] Inhibition was reversible upon removal of the NH_4Cl or monensin, suggesting that protection was not due to inhibition of toxin binding.

Additional studies demonstrated that pretreatment of macrophages with NH$_4$Cl also inhibited by >90% the rise in cellular cyclic AMP induced by edema toxin (PA + EF).[37] Similarly, Gordon et al.[38] looked at the effect of edema toxin (ET) on Chinese Hamster ovary (CHO) cells. Chloroquine and ammonium chloride inhibited intoxication of CHO cells with ET by over 99%. These results are consistent with the thesis that lethal toxin and edema toxin share a common internalization pathway mediated by PA, and that acidification of intracellular organelles plays a role in entry.

The fact that EF is an adenylate cyclase which is calmodulin dependent and has as its substrate ATP strongly suggests that translocation of EF across a vesicular membrane must occur to deliver the EF into the cytosol or to expose it to the cytosolic side of the membrane where it can interact with ATP. The presumed common mechanism of internalization of edema and lethal toxins, therefore, implies that LF must similarly be inserted into and perhaps translocated across a membrane to reach its putative target in the cytosol.

If the inhibition of toxin activity by amines and monensin is due to their ability to alkalinize intracellular vesicles, then it might be possible to overcome the block by lowering the pH, as has been demonstrated with diphtheria toxin,[39,40] some growth factors,[34] and viruses.[32] In these experiments,[22] macrophages were pretreated with NH$_4$Cl and then exposed to lethal toxin. The toxin was then removed by washing and cells were reincubated in the continued presence of NH$_4$Cl for another 60 min. The cells were then exposed for a 10-min period to media of varying pH before being returned to medium at pH 7.4. Under these conditions, the inhibition of toxicity by amines could be reversed if the pH was lowered to ≤ 6.0. Full toxicity was restored at pH 4.75 to 4.5, while no toxicity was observed if the pH remained neutral. A 1-min exposure to low pH was sufficient to reverse the inhibition. These results show that the protection by amines is due to an elevated vesicle pH and that lethal toxin-induced cytotoxicity is acid dependent. Similar observations were made with monensin-treated cells. Additionally, in experiments in nonphagocytic cells using edema toxin, Gordon et al.[38] also were able to overcome amine blockade by lowering the pH, although the exact details are not given.

The location of the lethal toxin at the time when exposure to medium of low pH can reverse the NH$_4$Cl-induced inhibition of cytotoxicity most probably is intracellular.[22] This conclusion is based upon the finding that antibody to LF, added in the cold immediately before low pH treatment, could not protect against the acid-induced cytotoxicity, implying that the LF was not accessible to the extracellular antibody. This interpretation is consistent with other reports showing that amines and monensin cause diphtheria toxin[39-41] and other protein ligands[42] to accumulate in acidic prelysosomal intracellular vesicles. Similar experiments demonstrating acidification induced translocation of diphtheria toxin from intracellular vesicles to the cytosol have been reported using transferrin-diphtheria toxin conjugates.[43] With diphtheria toxin itself, after an amine block and incubation at 37°C, some of the toxin is translocated from an intracellular location by low-pH treatment.[39] Thus, lethal and edema toxins appear similar to diphtheria toxin in that acidification can induce toxicity in amine-protected cells. These results suggest that, like the diphtheria toxin A fragment, LF, the A protein of lethal toxin, and EF, the A protein of edema toxin, are translocated to the cytosol to exert their biochemical effects and that experimentally induced translocation can occur from a prelysosomal compartment.

Similar approaches have been used to identify the intracellular site at which acidification normally induces lethal toxin toxicity as have been used to study other toxins;[44-47] these approaches include examining the time it takes lethal toxin to travel from the plasma membrane to the amine-inhibitable intracellular location and determining the effect of low temperature on toxicity. In these experiments, proteolytically activated PA + LF were prebound to cells at 4°C. The cells, washed free of unbound toxin, were reincubated at 37°C and NH$_4$Cl was added at various times to determine when toxicity could no longer be inhibited by elevation of intravesicular pH, implying the toxin had passed through the acidic site required for expression of toxicity. Even as early as 2 to 3 min after warming cells to 37°C, some toxin had already been exposed to an acidic environment. By 5 to 6 min, ≈ 50% of the maximal cytotoxicity had occurred and by 10 to 11 min, toxin had passed the amine-inhibitable step and 85% of the cells had been killed. Thus, it appears that lethal toxin is similar to diphtheria toxin[44] and pseudomonas exotoxin A,[47] but differs from modeccin[45,46] in that it passes through an acidic intracellular environment within a few minutes of internalization from the plasma membrane. In view of this finding and the knowledge that ligands in macrophages and other cell types require at least 20 min to reach lysosomes,[48] it is likely that lethal toxin encounters the acidic environment necessary for expression of toxicity at a prelysosomal level and implies that this may be the site of translocation. However, it has been noted with diphtheria toxin[44] and modeccin[46] that there is a lag between acidification and onset of protein synthesis inhibition suggesting translocation may occur sometime after the acidification step.

In other cell types, low temperature has been shown to interfere with movement of toxins,[44,45,47] ligands,[49,50] and viruses[51] from prelysosomal compartments to lysosomes. We have found that incubation of macrophages at 15°C protected them against lethal toxin over the 50-fold concentration range tested.[52] This is consistent with, but does not prove, that anthrax lethal toxin exerts its effect during or after passage from a prelysosomal compartment to the lysosome. The effect of low temperature on other cellular processes involved in toxicity remains to be determined. For example, as discussed below, although low temperature does not appear to inhibit the acidification step itself, it is not known whether anthrax toxin can translocate across membranes at low temperature as has been reported for diphtheria toxin.[40] Furthermore, no information is available as to whether low temperature interferes with the postulated but unknown enzymatic activity of LF. Experiments using edema toxin may be helpful in this regard.

As other workers have pointed out,[31,44] that, because of the very low number of toxin molecules which may be sufficient to intoxicate cells, there is considerable difficulty in interpreting experiments correlating effects of amines and low temperature on toxicity with measurements of the bulk flow of toxins. For example, it has been reported that incubation at 15°C does not block the toxicity of diphtheria toxin. This finding has been cited as evidence that toxin enters the cytosol from a prelysosomal vesicle; that there is no detectable lysosomal breakdown of diphtheria toxin at the low temperature implies that toxin does not reach the lysosome. However, it takes 100-fold more toxin to produce comparable toxicity at 15°C than at 37°C. This means that only 1% of the toxin passing the low temperature block to the lysosome could account for the observed toxicity. This quantity of toxin is beyond the limits of detectability using current methodologies. Thus definitive interpretations of the location of translocation are not possible.[44]

The observations of the inhibitory effect of low temperature on lethal toxin activity did allow us to examine the temporal relationship between the intracellular step which is amine sensitive and that which is inhibited by low temperature. Cells with toxin prebound at 4°C were incubated at 15°C for 8 h to block cytotoxicity. The cells were then incubated in the presence or absence of NH_4Cl and shifted to 37°C. Under these conditions, after release from the low temperature block, amines could no longer inhibit cytotoxicity. This implies that the cold temperature block occurs after toxin has passed through the amine inhibitable step, a situation similar to that reported for diphtheria toxin,[44,45] pseudomonas exotoxin A,[47] and modeccin.[45,46] Further, it suggests that low temperature does not interfere with the acidification process itself.

From this discussion, it is clear that information concerning intracellular trafficking of anthrax toxins is just beginning to develop. It will be of interest in the future to study naturally occurring or mutant cells which are resistant to the lethal and edema toxins and to examine the effects of anthrax toxins on the acidification defective mutants which have been reported.[53-57] Finally, the use of altered toxins and of chimeric toxins composed of anthrax toxin components may help us to understand the intracellular trafficking and mechanism of action of these toxins.

REFERENCES

1. **Lincoln, R. E. and Fish, D. C.,** Anthrax toxin, in *Microbial Toxins,* Vol. 3, Montie, T. C., Kadis, S., and Ajl, S. J., Eds., Academic Press, New York, 1970, 362.
2. **Stephen, J.,** Anthrax toxin, *Pharmacol. Ther.,* 12, 501, 1981.
3. **Gill, D. M.,** Seven toxic peptides that cross cell membranes, in *Bacterial Toxins and Cell Membranes,* Jeljaszewicz, J. and Wadstrom, T., Eds., Academic Press, New York, 1978, 291.
4. **Leppla, S. H.,** Anthrax toxin edema factor: A bacterial adenylate cyclase that increases cyclic AMP concentrations in eukaryotic cells, *Proc. Natl. Acad. Sci. U.S.A.,* 79, 3162, 1982.
5. **Woodin, A. M.,** Staphylococcal leukocidin, in *Microbial Toxins,* Vol. 3, Montie, T. C., Kadis, S., and Ajl, S. J., Eds., Academic Press, New York, 1970, 327.
6. **Iwasaki, M., Ohishi, L., and Sakaguchi, G.,** Evidence that Botulinum C_2 toxin has two dissimilar components, *Infect. Immun.,* 29, 390, 1980.
7. **Mikesell, P., Ivins, B. E., Ristroph, J. D., and Dreier, T. M.,** Evidence for plasmid-mediated toxin production in *Bacillus anthracis, Infect. Immun.,* 39, 371, 1983.
8. **Vodkin, M. and Leppla, S. H.,** Cloning of the protective antigen gene of *Bacillus anthracis, Cell,* 34, 693, 1983.
9. **Robertson, D. L. and Leppla, S. H.,** Molecular cloning and expression in *Escherichia coli* of the lethal factor gene of *Bacillus anthracis, Gene,* 44, 71, 1986.
10. **Tippetts, M. T., Robertson, D. L., and Leavitt, R.,** Molecular cloning and characterization of the *Bacillus anthracis* edema factor gene, presented at American Society for Microbiology, Atlanta, March 1 to 6, 1987, 29.
11. **Mock, M., Glaser, P., Danchin, A., and Ullman, A.,** Cloning of *Bacillus anthracis* adenylate cyclase gene and expression in *Escherichia coli,* presented at Centenary Symposium, Institut Pasteur, Paris, October 5 to 9, 1987, 123.
12. **Welkos, S. L., Lowe, J. R., Eden-McCutchan, F., Vodkin, M., Leppla, S. H., and Schmidt, J. J.,** personal communication, 1988.
13. **Robertson, D. L. and Leppla, S. H.,** personal communication, 1988.
14. **Stanley, J. L. and Smith, H.,** Purification of factor I and recognition of a third factor of the anthrax toxin, *J. Gen Microbiol.,* 26, 49, 1961.
15. **Beall, F. A., Taylor, M. J., and Thorne, C. B.,** Rapid lethal effect in rats of a third component found upon fractionating the toxin of *Bacillus anthracis, J. Bacteriol.,* 83, 1274, 1962.
16. **Molnar, D. M. and Altenbern, R. A.,** Alterations in the biological activity of protective antigen of *Bacillus anthracis* toxin, *Proc. Soc. Exp. Biol.,* 114, 294, 1963.
17. **Gladstone, G. P.,** Immunity to anthrax: protective antigen present in cell-free culture filtrates, *Br. J. Exp. Pathol.,* 27, 394, 1946.
18. **Hambleton, P., Carman, J. A., and Melling, J.,** Anthrax: the disease in relation to vaccines, *Vaccine,* 2, 125, 1984.
19. **Leppla, S. H.,** *Bacillus anthracis* calmodulin-dependent adenylate cyclase: chemical and enzymatic properties and interactions with eucaryotic cells, in *Advances in Cyclic Nucleotide and Protein Phosphorylation Research,* Vol. 17, Greengard, P. et al., Eds., Raven Press, New York, 1984, 189.
20. **Friedlander, A., White, J. D., Leppla, S., Tobery, S., and Merrill, P.,** Effect of bacterial adenylate cyclase on phagosome-lysosome function, *Fed. Proc.,* 42, 866, 1983.
21. **O'Brien, J., Friedlander, A., Dreier, T., Ezzell, J., and Leppla, S.,** Effect of anthrax toxin components on human neutrophils, *Infect. Immun.,* 47, 306, 1985.
22. **Friedlander, A. M.,** Macrophages are sensitive to anthrax lethal toxin through an acid-dependent process, *J. Biol. Chem.,* 261, 7123, 1986.
23. **Friedlander, A. M.,** unpublished data, 1985.
24. **Leppla, S. H., Ivins, B. E., and Ezzell, J. W., Jr.,** Anthrax Toxin in *Microbiology — 1985,* Lieve, L., Ed., American Society for Microbiology, Washington, D. C., 1985, 63.
25. **Wrobel, C., Youle, R., Leppla, S., and Friedlander, A.,** unpublished data, 1988.

26. **Ezzell, J. W., Ivins, B. E., and Leppla, S. H.**, Immunoelectrophoretic analysis, toxicity, and kinetics of *in vitro* production of the protective antigen and lethal factor components of *Bacillus anthracis* toxin, *Infect. Immun.*, 45, 761, 1984.

27. **Little, S. F., Leppla, S., and Cora, E.**, Production and characterization of monoclonal antibodies to the protective antigen component of *Bacillus anthracis* toxin, *Infect. Immun.*, 56, 1807, 1988.

28. **Friedlander, A., Leppla, S. and Cora, E.**, unpublished data, 1987.

29. **Leppla, S., Friedlander, A., and Cora, E.**, Proteolytic activation of anthrax toxin bound to cellular receptors, presented at Third European Workshop on Bacterial Toxins, Überlingen West Germany, June 28 to July 3, 1987.

30. **Middlebrook, J. L. and Dorland, R. B.**, Bacterial toxins: cellular mechanisms of action, *Microbiol. Rev.*, 48, 199, 1984.

31. **Olsnes, S. and Sandvig, K.**, Entry of polypeptide toxins into animal cells, in *Endocytosis*, Pastan, I. and Willingham, M. C., Eds., Plenum Press, New York, 1985, 195.

32. **Helenius, A., Kartenbeck, J., Simons, K., and Fries, E.**, On the entry of Semliki Forest virus in BHK-21 cells, *J. Cell Biol.*, 84, 404, 1980.

33. **Madshus, I. H., Olsnes, S., and Sandvig, K.**, Mechanism of entry into the cytosol of Poliovirus type 1: requirement for low pH, *J. Cell. Biol.*, 98, 1194, 1984.

34. **King, A. C. and Cuatrecasas, P.**, Exposure of cells to an acidic environment reverses the inhibition by methylamine of the mitogenic response to epidermal growth factor, *Biochem. Biophys. Res. Commun.*, 106, 479, 1982.

35. **Ohkuma, S. and Poole, B.**, Fluorescence probe measurement of the intralysosomal pH in living cells and the perturbation of pH by various agents, *Proc. Natl. Acad. Sci. U.S.A.*, 75, 3327, 1978.

36. **Maxfield, F. R.**, Weak bases and ionophores rapidly and reversibly raise the pH of endocytic vesicles in cultured mouse fibroblasts, *J. Cell Biol.*, 95, 676, 1982.

37. **Friedlander, A. M.**, unpublished data, 1985.

38. **Gordon, V. M., Leppla, S. H., and Hewlett, E. L.**, Inhibitors of receptor-mediated endocytosis block the entry of *Bacillus anthracis* adenylate cyclase toxin, but not *Bordetella pertussis* adenylate cyclase toxin, *Infect. Immun.*, 56, 1066, 1988.

39. **Sandvig, K. and Olsnes, S.**, Diphtheria toxin entry into cells is facilitated by low pH, *J. Cell Biol.*, 87, 828, 1980.

40. **Draper, R. K. and Simon, M. I.**, The entry of diphtheria toxin into the mammalian cell cytoplasm: evidence for lysosomal involvement, *J. Cell Biol.*, 87, 849, 1980.

41. **Marnell, M. H., Stookey, M., and Draper, R. H.**, Monensin blocks the transport of diphtheria toxin to the cell cytoplasm, *J. Cell Biol.*, 93, 57, 1982.

42. **Merion, M. and Sly, W. S.**, The role of intermediate vesicles in the adsorptive endocytosis and transport of ligand to lysosomes by human fibroblasts, *J. Cell Biol.*, 96, 644, 1983.

43. **O'Keefe, D. O. and Draper, R. K.**, Characterization of a transferrin-diphtheria toxin conjugate, *J. Biol. Chem.*, 260, 932, 1985.

44. **Marnell, M. H., Shia, S-P., Stookey, M., and Draper, R. K.**, Evidence for penetration of diphtheria toxin to the cytosol through a prelysosomal membrane, *Infect. Immun.*, 44, 145, 1984.

45. **Sandvig, S., Sundan, A., and Olsnes, S.**, Evidence that diphtheria toxin and modeccin enter the cytosol from different vesicular compartments, *J. Cell Biol.*, 98, 963, 1984.

46. **Draper, R. K., O'Keefe, D. O., Stookey, M., and Graves, J.**, Identification of a cold-sensitive step in the mechanism of modeccin action, *J. Biol. Chem.*, 259, 4083, 1984.

47. **Morris, R. E. and Saelinger, C. B.**, Reduced temperature alters pseudomonas exotoxin A entry into the mouse LM cell, *Infect. Immun.*, 52, 445, 1986.

48. **Wileman, T., Harding, C., and Stahl, P.**, Receptor-mediated endocytosis, *Biochem. J.*, 232, 1, 1985.

49. **Dunn, W. A., Hubbard, A. L., and Aronson, N. N., Jr.**, Low temperature selectively inhibits fusion between pinocytic vesicles and lysosomes during heterophagy of ^{125}I-asialofetuin by the perfused rat liver, *J. Biol. Chem.*, 255, 5971, 1980.

50. **Wolkoff, A. W., Klausner, R. D., Ashwell, G., and Harford, J.**, Intracellular segregation of asialoglycoproteins and their receptor: a prelysosomal event subsequent to dissociation of the ligand-receptor complex, *J. Cell Biol.* 98, 375, 1984.

51. **Marsh, M., Bolzau, E., and Helenius, A.**, Penetration of Semliki Forest virus from acidic prelysosomal vacuoles, *Cell*, 32, 931, 1983.

52. **Friedlander, A. M.**, unpublished data, 1988.

53. **Merion, M., Schlesinger, P., Brooks, R. M., Moehring, J. M., Moehring, T. J., and Sly, W. S.**, Defective acidification of endosomes in Chinese hamster ovary cell mutants "cross-resistant" to toxins and viruses, *Proc. Natl. Acad. Sci. U.S.A.*, 80, 5315, 1983.

54. **Robbins, A. R., Oliver, C., Bateman, J. L., Krag, S. S., Galloway, C. J., and Mellman, I.**, A single mutation in Chinese hamster ovary cells impairs both Golgi and endosomal functions, *J. Cell Biol.*, 99, 1296, 1984.

55. **Marnell, M. H., Mathis, L. S., Stookey, M., Shia, S-P., Stone, D. K., and Draper, R. K.,** A Chinese hamster ovary cell mutant with a heat-sensitive, conditional-lethal defect in vacuolar function, *J. Cell Biol.,* 99, 1907, 1984.

56. **Roff, C. F., Fuchs, R., Mellman, I., and Robbins, A. R.,** Chinese hamster ovary cell mutants with temperature-sensitive defects in endocytosis. I. Loss of function on shifting to the nonpermissive temperature, *J. Cell Biol.,* 103, 2283, 1986.

57. **Cain, C. C. and Murphy, R. F.,** A chloroquine-resistant Swiss 3T3 cell line with a defect in late endocytic acidification, *J. Cell Biol.,* 106, 269, 1988.

Chapter 8

INTRACELLULAR TRAFFICKING OF RICIN IMMUNOTOXINS: THE ROLE OF RICIN B-CHAIN

Jill Marie Manske and Daniel A. Vallera

TABLE OF CONTENTS

I. INTRODUCTION

Immunotoxins (IT) formed by covalently linking specific antibodies to enzymatic toxins show promise for treatment of human diseases. At the University of Minnesota, we have transplanted over 60 patients using bone marrow treated with intact ricin IT.[1] The delivery of cytotoxic signals to target cells using exquisitely selective monoclonal antibodies (MoAb) is straightforward in theory. However, despite the use of similar antibodies, IT made with different toxins may vary in their effectiveness.[2,3] One variable that could influence the efficacy of these reagents is the ability of IT to traffic within cells. This review focuses on the intracellular trafficking of IT to the lysosomal compartment using subcellular fractionation techniques. More specifically, we address the role of ricin B-chain in directing anti-CD5 IT to the lysosomes of CEM cells. CD5 is an antigenic determinant expressed on the surface of normal and malignant T cells,[4] and CEM is a CD5-positive T-lineage acute lymphoblastic leukemia cell line. The selectivity and potency of anti-CD5 IT have been previously established.[5-20] This review compares the trafficking patterns of anti-CD5 IT made with either ricin toxin A-chain (RTA) or intact ricin containing both RTA and B-chain.

II. RICIN IMMUNOTOXINS

A. RICIN

Although many toxins are currently under investigation, the intact toxin ricin has been used extensively for the preparation of effective IT. Ricin is a plant lectin derived from the seeds of the castor bean plant *Ricinus communis*. Ricin is synthesized as a single molecule which is subsequently processed into two 30-kDa chains, an A-chain and a B-chain, each with distinct functions. The B-chain of ricin is linked to the A-chain by a single disulfide bridge at a cysteine residue in amino acid position 257 of RTA.[6] Ricin B-chain has lectin properties and binds to carbohydrates with terminal galactose residues. Thus, galactose and lactose are efficient inhibitors of binding because of their competition with the receptor for the toxin binding sites.[21] Ricin binds to different oligosaccharides terminating in Gal-GlcNAC-Man,[22] and some Gal-GalNAC residues have also been observed to bind to the toxin. It is, therefore, clear that ricin is able to bind to a large number of different oligosaccharides at the cell surface. These sugars may in turn be linked to many glycoproteins or glycolipids, allowing ricin to bind to numerous cell surface molecules. It is, therefore, not surprising that the number of ricin binding sites on cells has been demonstrated to be quite high, 3×10^7 in the case of HeLa cells.[23] While the B-chain of ricin can efficiently bind to the cell surface, it has no toxic properties of its own.

The A-chain of ricin is responsible for the potent toxic activity of ricin. RTA is a potent catalytic enzyme that inhibits protein synthesis by inactivating 60S ribosomes of eukaryotic cells. No energy or cofactors are required, and kinetic experiments have shown that solutions of pure RTA can inactivate ribosomes at a rate of 1500 ribosomes per minute per toxin A-chain.[24] Investigation indicates that a single molecule of RTA present in the cytosol is sufficient to kill a cell.[25] RTA has recently been shown to act directly on 28S rRNA.[26] These experiments indicated that RTA inactivated eukaryotic ribosomes by a specific modification of the adenine[4324] residue of 28S rRNA. The nature of the modification has yet to be established; however, RTA probably acts as a specific N-glycosidase to remove adenine from rRNA.[26] The enzymatic activity of RTA is only expressed when the disulfide bonds between ricin A-chain and B-chain are reduced and A-chain is liberated.[27] The exact intracellular site at which this occurs is unknown, but it presumably must occur before A-chain can gain access to the cytosol.

Toxicity of ricin in tissue culture showed that ricin killed cells in concentrations of 1 ng/ml, with the earliest demonstrable effect being inhibition of protein synthesis.[28] DNA synthesis decreased somewhat later, and still later, RNA synthesis was inhibited.[28-30] There was no early effect of ricin on energy metabolism or oxidative phosphorylation. After addition of ricin to cells, a temperature-dependent lag time before inhibition of protein synthesis is apparent. The lag decreases with increasing toxin concentration, but even at high concentrations, it is still 20 to 30 min.[31] This is consistent with the view that endocytosis is an obligatory step for ricin toxicity.[32] From these observations, it is clear that the A- and B-chains of ricin work together to form a highly toxic molecule with efficient mechanisms for binding, entering, and killing eukaryotic cells.

B. INTACT RICIN IMMUNOTOXINS

Potent IT can be formed by coupling ricin to MoAb that are chosen based on the subset of cells to be killed. When the native binding site of ricin B-chain is blocked by lactose, toxin can be directed via antibody to a select cell population, leaving nontarget cells unaffected. There are several methods available by which ricin can be conjugated to MoAb. For conjugation to intact ricin, the heterobifunctional cross-linking reagent *M*-maleimidobenzoyl-*N*-hydroxysuccinimide ester (MBS) has been used successfully.[33] By this method, MoAb and toxin are linked in a two-step process in which *N*-hydroxysuccinimide reacts with amino groups present on the toxin. The maleimide residue can then combine with thiol groups on the MoAb, and toxin is thus linked to MoAb by a stable, covalent thioether bond.[33]

C. RICIN A-CHAIN IMMUNOTOXINS

Ricin IT can also be synthesized using only RTA. Immunotoxins formed in this manner do not require a lactose blockade since the native binding site of Ricin B-chain is not present. Antibody and RTA can be cross-linked using *N*-succinimidyl-3-(2-pyridyldithio)-propionate (SPDP). The MoAb is derivatized with SPDP, followed by conjugation to available sulfhydryl groups on the reduced toxin A-chain.[34] This results in the formation of a reducible disulfide bond between the antibody and toxin moiety. If the disulfide bond is replaced by a more stable one, such as a thioether bond, the cytotoxicity of the resulting IT is markedly reduced.[35] It is presumed that the disulfide bond between RTA and MoAb must be cleaved before RTA translocates into the cytoplasm. Unfortunately, RTA-IT can be as much as 100-fold less toxic than intact ricin IT.[5,36] This observation has led to the belief that ricin B-chain serves a function in additon to binding.

D. FUNCTION OF RICIN B-CHAIN

In addition to a role in binding, a translocation enhancing effect has been reported for ricin B-chain.[2] The addition of purified ricin B-chain can accelerate the cytotoxic activity of A-chain IT, without increasing the amount of A-chain bound to cells. Other studies have shown that the toxicity of A-chain IT can be enhanced *in vitro* or *in vivo* by the addition of free B-chain or B-chain coupled to a MoAb recognizing the same determinant as the A-chain IT.[37-39] The presence of B-chain somehow increases the rate of A-chain transport to the ribosomes and augments the toxicity of RTA. The region of the B-chain responsible for facilitating A-chain activity has not yet been identified.[6] However, a conformational change caused by incorporation of the hydrophobic regions of ricin B-chain into the hydrophobic region of membranes may induce translocation of RTA across the membrane and into the cytosol.[40]

Studies are underway to determine whether the binding activity and A-chain enhancing activity of ricin B-chain are separable. Investigators have shown that chloramine T-mediated oxidation and iodination of B-chains, either before or after coupling to antibody, reduces the binding activity of B-chain 100-fold while reducing the potentiating ability only 2- to 8-fold.[41] These findings support the view that the binding site is not essential for B-chain enhancement of A-chain activity. However, other investigators have implicated the galactose binding site of ricin B-chain as a critical factor for toxin entry into the cytosol prior to protein synthesis inhibition.[42] When acetylation was used to inhibit the galactose binding ability of ricin B-chain, a 90% decrease in toxicity of the IT was observed,[42] suggesting that binding and facilitation of A-chain activity may not be separable. Other studies have shown that acetylricin IT are tenfold less toxic than the nonacetylated counterparts.[43] These data argue that the binding and enhancing functions of ricin B-chain may not be separated. Further studies into the exact mechanism of B-chain enhancement will be required to determine if these functions are indeed separable.

E. ANTI-CD5 RICIN IMMUNOTOXINS

Effective IT have been formed by linking ricin to the MoAb T101, which recognizes the CD5 determinant.[7-18] The MoAb T101 was first described by Royston et al. in 1979.[4] The antibody was produced by immunizing mice with a human leukemia cell line 8402.[44] The MoAb, designated T101, was found to bind all human cells of thymic origin, including peripheral blood T lymphocytes, thymus cells, and malignant T cell leukemias and lymphomas.[4] Reactivity of the MoAb was later found to include surface immunoglobulin-positive chronic lymphocytic leukemia cells,[45] a small number of B-cells in adult lymph nodes,[46,47] and a subpopulation of fetal B-cells.[48] Since the MoAb reacts with only a small percentage of B-cells, it is classified as a pan T cell antibody. The T101 antibody was found to precipitate a 67,000-mol wt glycoprotein[4,49] was later termed CD5.[50] The CD5 molecule consists of a 58,000-mol wt protein backbone to which carbohydrates are added.[49] High-mannose sugars are added to asparagine residues during synthesis; then within 20 min, these high-mannose sugars are converted to complex carbohydrates and the fully processed 67,000-mol wt glycoprotein appears at the cell surface within 30 min after synthesis.[49]

Until recently, the function of the CD5 antigen remained unknown. The antigen is now thought to be involved in the activation and proliferation of T cells.[51] In 1985, Ledbetter et al. showed that anti-CD5 MoAb could augment and sustain the proliferative response of activated T cells.[51] The anti-CD5 antibodies could not induce activation of T cells by themselves. The CD5 receptor was, therefore, assumed to function as a receptor for "second signals" in T cells. Subsequent work by Ceuppens and Baroja[52] supported the observation that antibodies to CD5 could serve as a second signal in T cell activation. They found that MoAb to CD5 could provide an accessory signal for T cell activation, replacing signals otherwise provided by accessory cells and/or interleukin 1.[52]

Anti-CD5 MoAb conjugated to ricin have been shown to be highly effective IT.[7-18] The conjugates are cytotoxic to normal and malignant cells bearing the CD5 determinant, while sparing antigen-negative cells. Anti-CD5 have been used for the eradication of T-cells from bone marrow grafts,[10,11,14,53-57] for specific killing of T-leukemia cells,[7,9,12-14,16,55,58,59] and in a number of tumor model systems.[5,17,18]

III. CLINICAL APPLICATIONS OF RICIN IMMUNOTOXINS

A. AUTOLOGOUS BONE MARROW TRANSPLANTATION

Based on the ability of anti-CD5-ricin and other IT to efficiently kill subsets of cells, IT may be useful for the deletion of particular cell types in suspensions of bone marrow. In T101-ricin1984, Stong et al. showed that anti-CD5-ricin had marked and specific cytotoxicity against human T-leukemia cells.[9] In an experiment where human leukemia cells from a clonogenic cell line were mixed with irradiated bone marrow, anti-CD5 ricin eliminated >99.99% of the leukemia cells from human bone marrow with minimal effect on multipotent stem cells. These studies provided the rationale for the application of anti-CD5-IT for the *ex-vivo* eradication of T cell leukemia and lymphoma cells from bone marrow prior to autologous transplantation.[1] Bone marrow transplantation (BMT) is used as an adjuvant in treatment of cancers which are susceptible to irradiation and/or chemotherapy.[60,61] In the case of autologous BMT, marrow obtained from patients in remission is frozen and stored. If the patient relapses, the tumor is eradicated using irradiation and/or chemotherapy. The patient is then rescued by infusion of his own bone marrow. However, it may be necessary to purge the marrow of latent cancer cells prior to infusion. At the University of Minnesota, patients with T-lineage acute lymphoblastic leukemia (T-ALL) have received IT-treated autologous BMT.[59,62] Indications from these studies are that IT represent a safe, specific, simple method for purgation of autologous bone marrow. The value of the procedure remains to be tested in randomized clinical trials.[34] Similar studies have been undertaken using anti-CD5-RTA plus the addition of the potentiator ammonium chloride.[12,14]

B. ALLOGENEIC BONE MARROW TRANSPLANTATION

Anti-T-cell IT including anti-CD5-ricin may be useful in allogeneic BMT for the *ex vivo* elimination of mature T cells from donor marrow in order to prevent graft-vs.-host-disease (GVHD).[14,34,53-57,63,64] The abnormal hematopoietic cells of patients with diseases, such as aplastic anemia, leukemia, or immunodeficiency disease are ablated by chemoradiotherapy prior to transplant. A bone marrow graft from a human leukocyte antigen (HLA) matched donor is then infused into the immunosuppressed recipient. Immunocompetent T cells from the donor graft can recognize antigens present on the cells of the recipient and cause GVHD, a serious pathological reaction to recipient tissue.[34] Prevention of GVHD by IT has been reported in murine systems[54,63] and clinical trials.[53-55]

C. *IN VIVO* USE OF IMMUNOTOXINS

Current research in the field of IT includes the *in vivo* application of these reagents. The requirements for use of IT *in vivo* are more stringent than those for *ex vivo* applications. However, several researchers have demonstrated antitumor effects mediated by IT in murine systems.[65-69] Although the obstacles to the *in vivo* use of IT are considerable, these studies suggest promise for the use of IT in the treatment of localized and/or inoperable malignancies.

IV. ENDOCYTOSIS

One of the fundamental properties of living cells is their ability to sense and respond to their external environment. This is accomplished, in part, by bringing biologically important molecules from the cell exterior to the interior. Small molecules such as ions, amino acids, and sugars simply flow through or are pumped through the membrane by specific transport systems. Larger molecules, such as plasma proteins, hormones, toxins, and viruses must be ingested by another route.[70] This process is termed endocytosis. Basically, during the process of endocytosis, a patch of plasma membrane surrounds the material to be ingested. The material is then brought into the cell enclosed in a vesicle derived from the plasma membrane.[71]

Three different processes involving three different types of organelles carry out endocytic events. The engulfment of large particles by specialized cells is termed phagocytosis, and the organelle formed is a phagosome.[71] Pinocytosis is an endocytic process that results in the nonspecific uptake of extracellular fluid. The uptake of large bubbles of extracellular medium is termed macropinocytosis, and the organelle formed is a macropinosome.[72,73] The pathways of phagocytosis and pinocytosis result in the nonspecific uptake of molecules and fluid from the extracellular environment. The third process by which molecules can enter cells is termed receptor-mediated endocytosis, and is, by contrast, exquisitely specific.

A. RECEPTOR-MEDIATED ENDOCYTOSIS

The concept of receptor-mediated endocytosis was formulated in 1974 to explain the observation that regulation of cellular cholesterol metabolism depended on cell surface binding, internalization, and intracellular degradation of plasma low density lipoprotein (LDL).[74] Cholesterol is stored as LDL particles. Cells acquire cholesterol for synthesis of membranes by the receptor-mediated endocytosis of circulating LDL.[75] When cells require cholesterol, they synthesize a cell surface receptor that binds and internalizes LDL. The pathway of LDL endocytosis is similar to that of many other ligands.

The initial occurrence in receptor-mediated endocytosis is binding of a ligand to its specific receptor. The receptor/ligand complexes then slide laterally into clathrin-coated pits. Clathrin-coated pits are the unique plasma membrane site at which ligand/receptor complexes destined for endocytosis are accumulated and concentrated.[71] Coated pits were first observed by Roth and Porter in the plasma membrane of mosquito oocytes[76] and were later shown to occur in invertebrate and vertebrate cell types, including protozoa, yeast, hydra, and higher animals.[71,77] Receptors can either be accumulated in coated pits in a constituative manner without bound ligand (e.g., LDL receptor),[75] or they may concentrate within pits subsequent to binding of a specific ligand (e.g., epidermal growth factor [EGF] receptor).[78] Within minutes of receptors sliding into coated pits, the pits invaginate into the cell and pinch off to form coated endocytic vesicles. These vesicles shed their clathrin coats and fuse with one another to form vesicles known as endosomes[79,80] or receptosomes.[70,81]

The receptosomes are acidified by ATP-driven proton pumps[82-84] to a pH of approximately 4.5 to 5.0.[82] These vesicles do not contain significant amounts of functional hydrolytic enzymes[81,82,85] and, therefore, ligands and receptors are not extensively degraded or grossly modified within this compartment.[71] Receptosomes move by saltatory motion along tracks of microtubules[81] until they come in contact with the elements of the *trans*-Golgi. They then appear to fuse with the Golgi system, delivering their contents into the lumen of the Golgi.[78,86] It is in the *trans*-Golgi that sorting of ligand and receptor occurs. Also, the ultimate fate of the ligand and receptor may be determined in this organelle.[71,86] The possible fates are as follows: the ligand and receptor can both be returned to the cell surface where the ligand is released and the receptor reutilized (e.g., transferrin); the ligand may be directed to the lysosomes for degradation while the receptor is recycled (e.g., alpha-2-macroglobulin and LDL); or the ligand and its receptor can both be sent on to the lysosomes to be degraded (e.g., EGF and its receptor). Thus, the Golgi is an important "cross-roads" for directing intracellular traffic. It has been found that the tubules of the *trans*-Golgi are dotted with small clathrin-coated pits.[71] These small pits may play an important role in the sorting process since ligands destined for the lysosomes have been found concentrated in these coated pits[71,78] while ligands to be returned to the cell surface have not been found in these structures.[78] The final organelle in the endocytic pathway is the lysosome. Lysosomes are rich in proteolytic enzymes which are active in the low intralysosomal pH. Due to these enzymes, proteins delivered to the lysosomes are usually degraded completely to amino acids.[75]

In summary, the structural components involved in receptor-mediated endocytosis are (1) the plasma membrane, (2) clathrin-coated pits, (3) receptosomes, (4) Golgi, and (5) lysosomes. The routes available for internalized receptor/ligand complexes are (1) receptor is recycled, ligand is degraded, (2) receptor and ligand are both recycled, and (3) receptor and ligand are both degraded.[80]

B. ENDOCYTOSIS OF RICIN

Several laboratories have studied by electron microscopy (EM) the uptake of toxins linked to horseradish peroxidase (HRP), ferritin, and colloid gold.[21] A major problem with these studies is conjugate size. Also, when the conjugate is exposed to low pH and proteolytic enzymes after internalization, the label may be released and follow an intracellular route other than the route of the toxin. Even if the conjugate remains intact, its larger size compared to free toxin and its altered properties may affect the transport of the toxin.[21] Also, since ricin can bind to a variety of different cell surface molecules, the internalized toxin may travel through the intracellular compartments by different routes, only one of which may be relevant to intoxication. Finally, the most important event, how and where RTA is translocated through the limiting membrane, is unlikely to be visualized by these methods.

In spite of these limitations, ultrastructural studies have provided important information. Extensive EM studies have been carried out with ricin. Nicolson et al.[87,88] showed that ricin-ferritin complexes were bound in a disperse manner and exclusively at the cell surface at 4°C. With increasing time at 37°C, an increased fraction was internalized. The conjugate first appeared to cluster at the cell surface and was subsequently taken into endocytic vesicles. After 60 min, most of the toxin was present in these vesicles. The toxin was not observed to enter lysosomes. Also, certain ricin-resistant mutants were found to be less able to cluster and internalize ricin than were the parent cells.[21,89-92] Gonatas et al.[93-95] labeled ricin with HRP to study its uptake. When incubation was at 4°C, a continuous rim of the plasma membrane was stained. However, when cells were washed and incubated at 37°C, the plasma membrane became patchy. After 30 min, various degrees of staining appeared in vesicles. Clusters of stained vesicles were found adjacent to the cisternae of the Golgi. After 1 h, vesicles were found near the *trans*-Golgi, and at the edges of the cisternae. Even after 3 h at 37°C, some small patches of staining remained at the cell surface. However, most of the label was found in intracellular compartments, particularly in the Golgi cisternae. Notably, the uptake of ricin from the cell surface is much slower than the uptake of EGF and LDL, which within minutes undergo endocytosis via coated pits.[21] Clearly, endocytosis of ricin is quantitatively and qualitatively different from that of EGF and LDL.

After toxin is taken up by endocytosis, it may be transported to different compartments of the vesicular and tubular system in the cell before it eventually crosses the limiting membrane to gain access to the ribosomes. Internalized ricin appears to be able to enter the cytosol for hours after endocytosis.[96] Ricin is accumulated only to a low degree in lysosomes, and is degraded very slowly. Sandvig et al. showed that even after 2 h, about 90% of internalized ricin remained active.[97] Several researchers have reported ricin trafficking to the Golgi, and some have reported ricin binding to the cisternae of this organelle.[93,95,98,99] Recently, van Deurs et al. found that temperatures which allowed ricin internalization but excluded it from reaching the Golgi (18°C) abolished the ability of ricin to inhibit protein synthesis.[100] These data support a Golgi role in ricin toxicity, and suggest that transport to the Golgi may be important for intoxication.

C. ENDOCYTOSIS OF RICIN IMMUNOTOXINS

Although there is great interest in endocytosis and trafficking of ricin IT, very little is known about the entry mechanism of these reagents. As previously stated, the observed difference in toxicity between RTA IT and intact ricin IT has led to speculation that ricin B-chain serves an additional role in ricin activity beyond binding. A translocation-enhancing effect has been reported for B-chain.[2,39] It has also been proposed that ricin B-chain may protect A-chain from lysosomal degradation, since intact ricin is much more resistant to proteolysis *in vitro* than the separate subunits.[101] The role of B-chain in the internalization of anti-CD5 IT is undetermined.

Since lysosomal trafficking may be important to toxicity, we investigated the role of B-chain in lysosomal trafficking of IT, hypothesizing that intact ricin IT would be transferred to the lysosomes at a slower rate relative to RTA IT.[102] Slower trafficking could then allow for greater translocation of RTA into the cytosol before lysosomal delivery resulted in degradation of the protein. RTA IT was not as toxic to CEM cells as intact ricin IT (Figure 1). These differences were not attributed to differential binding or differences in modulation of the CD5 determinant as measured by FACS analysis (Table 1). The intracellular transport of anti-CD5 conjugated to intact ricin (anti-CD5-Rc) and anti-CD5-RTA was determined in

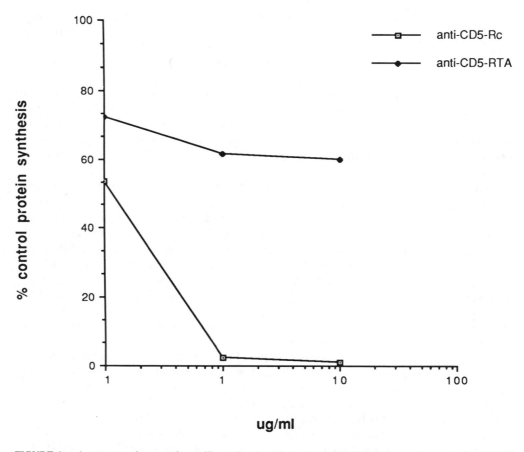

FIGURE 1.   A representative experiment illustrating the effect of anti-CD5 IT on the protein synthesis of CEM cells. Cells were pretreated with IT for 2 h. After three washes, cells were incubated for 24 h, and were pulsed with ³H-leucine. Control cultures were incubated in lactose without IT. Percent contol ³H-leucine incorporation is plotted against increasing IT concentration.

TABLE 1
Binding and Antigenic Modulation of Anti-CD5 MoAb and IT on CEM cells

	% Binding (4°C)[a]	% Modulation[b]
Anti-CD5	93.9	34
Anti-CD5-Rc	93	41
Anti-CD5-RTA	95.1	47

[a]   A representative experiment showing the binding of MoAb or IT to CEM cells. Cells were incubated with 1 µg/ml anti-CD5 or anti-CD5 IT for 30 min at 4°C. Binding was analyzed by indirect immunofluorescence using a FACS IV flow cytometer.

[b]   A representative experiment showing modulation of the CD5 determinant from the surface of CEM cells. Methodology has been previously reported.[15] Cells were incubated with MoAb or IT, washed, and warmed to 37°C for modulation. After 2 h, cells were incubated with saturating concentrations of anti-CD5 MoAb to bind any remaining CD5 antigen. Cells were then incubated with FITC-labeled goat-anti-mouse antibody. Cells were analyzed by FACS IV, and percent modulation was determined by comparison to nonmodulated controls.

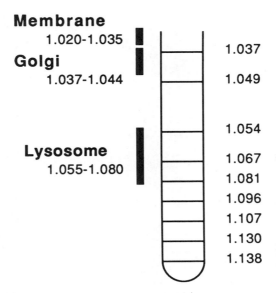

Membrane
1.020-1.035

Golgi
1.037-1.044

Lysosome
1.055-1.080

1.037

1.049

1.054

1.067

1.081

1.096

1.107

1.130

1.138

FIGURE 2. Illustration of a Percoll gradient obtained after ultracentrifugation of cellular organelles. Organelles are separated by density and the approximate densities of the plasma membrane, Golgi, and lysosomes are indicated.

the human malignant T cell line CEM using subcellular fractionation techniques. To analyze the intracellular location of radiolabeled anti-CD5-Rc and anti-CD5-RTA at various times after entry, the cells were homogenized and organelles fractionated on Percoll gradients. The *in situ* generated density gradient (Figure 2) was collected and fractions were assayed for radioactivity. A variety of enzymes were assayed in gradient fractions to determine the organelle distribution in CEM cells. The lysosomal marker enzyme, hexosamidase, consistently sedimented at the bottom of the gradient in fractions 14 to 20 (Figure 3).

Figure 4 shows data from a representative experiment illustrating the transfer of anti-CD5-RTA and anti-CD5-Rc to the lysosome containing fractions after 2 h at 37°C. Fractions from cells incubated at 4°C indicated a peak of radioactivity near the top of the gradient corresponding to the plasma membrane and/or Golgi fractions as indicated by enzyme analysis. The gradient profiles of anti-CD5-RTA showed co-sedimentation of radioactivity with the lysosome containing fractions after 2 h at 37°C. Anti-CD5-Rc at 37°C showed no transfer of radiolabeled reagent to the heavier fractions at any time points tested, suggesting that little or no anti-CD5-Rc was transferred to the lysosomes.

Transfer of ^{125}I-IT to lysosome containing fractions at 37°C was directly compared to values obtained at 4°C by *t*-test analysis (Table 2). These data show that by 2 h, anti-CD5-RTA co-sediments with the lysosomes ($p < 0.05$) and the amounts remain relatively constant from 2 to 4 h. When lysosomal fractions were pooled for cells incubated with ^{125}I-anti-CD5-Rc, the results showed that little or no detectable anti-CD5-Rc was trafficked to the lysosomes within 4 h at 37°C as compared to 4°C controls (Table 2).

A direct comparison of lysosomal associated radioactivity for anti-CD5-RTA and anti-CD5-Rc was also performed (Table 3). By 2 h, significantly more ^{125}I-labeled anti-CD5-RTA was observed co-sedimenting with the lysosomes as compared to anti-CD5-Rc. This difference was retained at 3 and 4 h. These data indicated that significantly more anti-CD5-RTA than anti-CD5-Rc co-sedimented with the lysosomes between 2 to 4 h of incubation at 37°C, and that the overall rate of trafficking to lysosome containing fractions was slower for anti-CD5 IT-containing B chain.

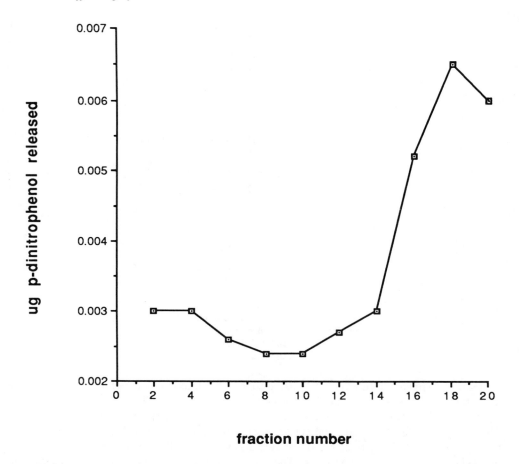

FIGURE 3.   Hexosamidase activity across the CEM equilibrium gradient. A homogenate was prepared and applied to a Percoll solution. After ultracentrifugation, the gradient was collected. Fractions were assayed for hexosamidase activity to locate lysosome-containing fractions.

V. ANTIGENIC MODULATION

Although not necessarily correlated with endocytosis, antigenic modulation by an antibody can provide information relative to the entry of antibody/antigen complexes, and, therefore, IT/antigen complexes. The phenomenon was first described in 1963 by Boyse and Old who coined the name antigenic modulation.[103] After an antibody binds to its specific receptor on the cell surface, there is often a redistribution and decrease or loss of expression of the antigen termed modulation. Modulation is energy dependent and occurs optimally at 37°C, is partially inhibited at 22°C, and is completely inhibited at 4°C.[104,105] It is also dependent on cellular metabolism and can be inhibited by compounds such as sodium azide. Conversely, modulation is not influenced by inhibitors of protein synthesis.[105]

An example of antigenic modulation can be found after treatment of B cells with anti-immunoglobulin sera. Following exposure of cells to antibody, surface immunoglobulins are rapidly redistributed on the B cell membrane. This ligand-induced redistribution occurs in several stages.[106] First, within minutes, the antigen is clustered in small patches. This patching is not dependent on cellular metabolism, but is influenced by both antibody valency and low temperatures. The second stage is capping. During this stage the antigen/antibody complexes are concentrated at one pole of the cell. This stage is energy dependent. The final stage is endocytosis and subsequent degradation of the internalized immunoglobulin.[105,106]

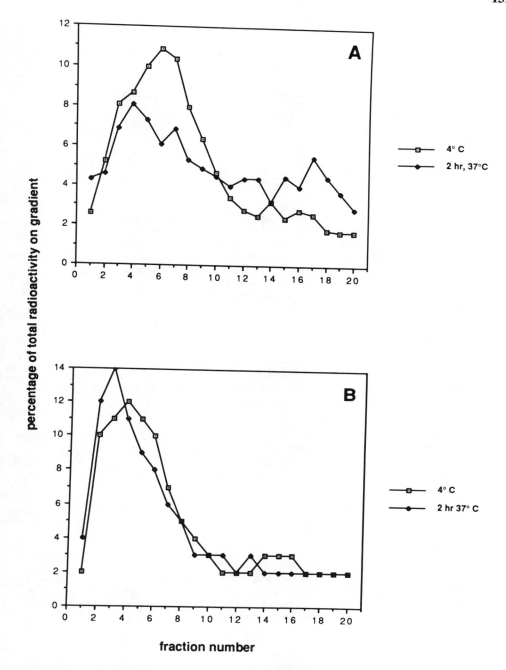

FIGURE 4.   Transfer of ^{125}I-anti-CD5-RTA (A) and ^{125}I-anti-CD5-Rc (B) between intracellular compartments after 2 h at 37°C as determined by equilibrium density centrifugation. In a representative experiment, CEM cells were incubated with 1 μg/ml IT at 4°C. Cells were washed and warmed to 37°C for 2 h. After homogenization and fractionation on Percoll gradients, fractions were harvested and the amount of radioactivity in each fraction was quantitated. A noninternalized 4°C control sample is plotted with each 37°C test sample.

TABLE 2
Comparison of Lysosomal Transfer of Anti-CD5
IT[a]

Time at 37°C	Significantly different from 4°C control	
	Anti-CD5-Rc	Anti-CD5-RTA
30 min	−	−
1 h	−	−
2 h	−	+
3 h	−	+
4 h	−	+

[a]   CEM cells were incubated with radiolabeled anti-CD5-Rc or anti-CD5-RTA for various time periods at 37°C. After homogenization and subcellular fractionation on Percoll gradients, aliquots were harvested and the lysosome containing fractions were pooled. The amount of radioactivity co-sedimenting with the lysosomes after the 37°C incubation was compared to a 4°C control by Student t-test analysis. Data are pooled from 5 to 7 experiments.

+, Yes (p <0.05).
−, No (not significant).

TABLE 3
Comparison of Lysosomal Transfer of Anti-CD5-IT in the Presence and Absence of Monensin[a]

Time at 37°C	Significantly different		
	Anti-CD5-Rc vs. anti-CD5-RTA	Anti-CD5-RTA + monensin vs. anti-CD5-RTA	Anti-CD5-RTA + monensin vs. anti-CD5-Rc
0	−	−	−
30 min	−	−	−
1 h	−	−	−
2 h	+	−	−
3 h	+	+	−
4 h	+	+	−

[a]   Cells were incubated with IT or IT plus 5 μM monensin at 37°C for various time periods. After homogenization and subcellular fractionation on Percoll gradients, aliquots were harvested and lysosome containing fractions were pooled. The amount of radioactivity co-sedimenting with the lysosomes after the 37°C incubation was compared by Student t-test analysis. Data are pooled from 5 to 7 experiments.

+, Yes (p <0.05).
−, No (not significant).

Differences in modulation are induced on various cells by different antibodies and antigens. For example, when the thymic leukemia (TL) antigen is modulated from TL cells by anti-TL antibodies, the TL antigen is not readily endocytosed, but is perhaps shed from the surface. Also, this antigen can be modulated by monovalent F(ab)′ fragments,[104,107] its modulation apparently not affected by valency.

The fate of a modulated antigen may depend on the biological function of the natural ligand for the receptor. Valency, as previously stated, may also play an important role. For example, Raso et al. have shown that anti-CALLA F(ab)₂-ricin is more effectively internalized and more toxic than F(ab)′-ricin prepared with the same antibody.[108] In another study, when transport of antibody to the Fc receptor on macrophages was studied, internalization of intact immunoglobulin resulted in degradation of both receptor and ligand. When the ligand was a monovalent F(ab)' fragment of the same antibody, the F(ab)'-receptor complex was recycled without degradation.[109] In the same experiment, multivalent F(ab)$_2$ preparations were as effective as intact immunoglobulin at mediating lysosomal transport, indicating that valency, and not the presence of intact Fc domains, was an important factor in determining receptor transport during endocytosis. These experiments underscore the necessity of testing individual antibodies to determine their modulatory behavior.

Several antibodies recognizing T cell determinants have been shown to induce modulation. Included among the numerous determinants showing modulation after exposure to their specific antibody are CD3, CD4, CD5, CD8, and the interleukin 2 receptor.[105] Modulation of CD5 by T101 and other antibodies has been studied by several investigators.[110-112] The CD5 antigen has been shown to modulate from both peripheral blood T cells and malignant cell lines,[110-112] and this modulation appears to result in internalization of the antibody/antigen complexes.[111] These results suggest that toxins bound to anti-CD5 MoAb would be effectively endocytosed during antigenic modulation induced by the antibody.

The modulation of anti-CD5 immunotoxins from the surface of normal and leukemic cells has been studied in order to determine whether the presence of toxin on antibody affects antigenic modulation.[15] In this study, anti-CD5-RTA modulated CD5 more efficiently than anti-CD5 conjugated to intact ricin, which modulated CD5 more efficiently than antibody alone (Table 1). Kinetic studies showed that maximum modulation of IT was reached within 3 h. Whether antigenic modulation plays a role in the entry and subsequent toxicity of IT is still unknown. However, these studies indicated that CD5 IT modulate like parent MoAb, and that the presence of toxin on antibody may actually enhance modulation and internalization of IT. In this study, radiolabeled reagents were employed to demonstrate that modulation of anti-CD5 IT involved internalization. As previously mentioned and as was shown in Figure 1, RTA IT are less toxic than intact ricin IT. When toxicity of the reagents was tested in protein synthesis inhibition assays, anti-CD5-RTA inhibited protein synthesis 23 to 43%, whereas anti-CD5 linked to intact ricin inhibited 99% of the protein synthesis of CEM cells. Taken together with Table 1, these data indicated that the lack of toxicity of T101-RTA chain was not attributable to its inability to bind or modulate.

VI. ROLE OF THE GOLGI IN IMMUNOTOXIN TRAFFICKING

As previously discussed in relation to endocytosis, increasing evidence supports a Golgi role in ricin or ricin IT trafficking.[93,95,98-100] Youle and Colombatti[113] showed that a hybridoma cell secreting antibodies against ricin is resistant to the toxin. A likely explanation of this observation is that the internalized ricin is inactivated by the anti-ricin antibody during transport through the Golgi complex.[113]

Some of the most convincing evidence of Golgi involvement in ricin trafficking has come from the use of potentiators of IT and ricin toxicity, such as monensin. Monensin is a well-characterized metabolite of *Streptomyces cinnamonensis* that binds Na$^+$, K$^+$, and protons.[114] It is lipophilic in nature and presumably can insert into all cellular membranes.[114] Carboxylic ionophores such as monensin are potent potentiators of IT activity.[19,115] Such agents can increase the specific toxicity of RTA IT from 4- to 50,000-fold depending on the target cells and surface antigen employed.[19] These compounds greatly increase the rate of inactivation of protein synthesis by RTA IT, which in some cases, approaches the rate observed with ricin.[19] The mechanism for this rate increase is unclear. Since carboxylic ionophores exchange monovalent cations across membranes, the exchange of K$^+$ for H$^+$ may increase the pH of acidic vesicles, such as lysosomes.[19] The potentiation could, therefore, be due to an inhibition of protein degradation.

Carboxylic ionophores and lysosomotropic amines also alter intracellular trafficking.[19,114,116] It is possible that these agents conduct IT to a compartment more suitable for translocation of RTA across limiting membranes. EM studies have shown that the addition of monensin does not inhibit entry of RTA IT, but decreases the amount of transfer to the lysosome.[20] It was concluded that monensin lengthened the time IT spent in nonlysosomal compartments, thereby prolonging the time allowed for RTA insertion into membranes of intracellular vesicles, and preventing its degradation.

The best-documented effect of monensin treatment is a radical slowing of intracellular transport of newly synthesized secretory proteins, proteoglycans, and plasma membrane glycoproteins.[114] Several normal, malignant, and virus-infected animal and plant cells have been studied. No exceptions are known.[114] The site of arrest of intracellular transport has been studied by several methods. In all cases, the principal site of arrest is the Golgi complex. This makes monensin unique.[114] Even at concentrations too low to affect the endosome or lysosomal pH, monensin still induces morphological changes in the Golgi and enhances ricin toxicity.[100] This suggests that alteration of intracellular transport of ricin IT can have a profound effect on toxicity.

To explore the possibility of a Golgi role in anti-CD5 IT trafficking, we employed monensin. When monensin was added to CEM cells incubated with anti-CD5-RTA, there was a dramatic and significant slowing of anti-CD5-RTA transfer to lysosome containing compartments (Table 3). The addition of monensin caused anti-CD5-RTA to traffic in a manner indistinguishable from anti-CD5-Rc (Table 3). Assuming that monensin blocked anti-CD5-trafficking at the Golgi, the observation that anti-CD5-Rc trafficked to the lysosomes similar to anti-CD5-RTA in the presence of monensin strongly suggests that anti-CD5-Rc containing B-chain traffics through the Golgi and is retained there relative to anti-CD5-RTA. Retention in the Golgi would delay transfer to lysosomes and may allow for RTA translocation from this organelle. From these studies, it was concluded that RTA IT were transferred to the lysosomes faster than intact ricin IT.[102]

Other groups have used EM to study the intracellular trafficking of ricin IT. The endocytosis of an IT prepared by linking RTA to a MoAb recognizing a carcinoma-associated antigen was studied.[40] These researchers found that the IT was internalized into the cell by two different pathways: one, via coated pits and coated vesicles, and another, via large enclosed invaginations. Both modes of endocytosis appeared to deliver the RTA IT to the lysosomes.[40] Internalization of IT was observed to be similar to that of MoAb alone. They also reported that during intracellular transport by either pathway, the IT remained intact until reaching the lysosomes. When monensin was used, there was more IT observed in endocytic vesicles, and less in lysosomes. These researchers also reported the visualization of IT in the cytosol, and reported that in areas of abundant endocytic vesicles both IT moieties (antibody and RTA) were found in the cytosol. In control experiments, intact ricin, but no unconjugated MoAb, was found in the cytosol after internalization.[40]

In a study where RTA was linked to an antibody to the transferrin receptor, Raso et al. showed that these IT followed the intracellular pathway of native, unconjugated transferrin.[117] These data suggest that highly effective IT might be chosen based on the intracellular pathway and endocytic behavior of the receptor recognized by the MoAb.

In a separate experiment using anti-CD5-RTA labeled with colloid gold, Carriere et al. showed that anti-CD5-RTA internalized via clathrin-coated pits, and eventually trafficked to the lysosomes of CEM cells.[20] These investigators also recognized a network of tubular elements near the Golgi from which anti-CD5-RTA was transferred to lysosomes. The addition of NH_4Cl or monensin, which increases the toxicity of anti-CD5-RTA, did not interfere with internalization into cells, but modified intracellular distribution of the labeled protein. Notably, trafficking of anti-CD5-RTA to the lysosomes was markedly slowed.[20]

In a different study, Press et al. used EM to demonstrate that both effective and ineffective

A-chain IT were endocytosed, but their rates of delivery to lysosomes differed.[118] The RTA-IT that were highly toxic were transferred to the lysosomes slower than the less toxic IT. In this study, IT were formed by linking RTA to various MoAb recognizing the CD5, CD3, and CD2 T-cell determinants. The authors concluded that there may be several distinct pathways for IT internalization and that the ability of RTA to inactivate ribosomes may depend on the specific membrane receptor involved in the binding of an IT, its route of internalization, and the rate of entry into lysosomes.[118] Our observation that anti-CD5-RTA was trafficked to lysosomes rapidly correlates with these observations. Press et al. hypothesized that pathways involving delayed fusion with lysosomes could allow more efficient A-chain penetration to the cytosol than pathways associated with prompt fusion. Although our methods differ, our data on trafficking of anti-CD5-RTA and anti-CD5-Rc support this hypothesis.

A possible model of IT intoxication, based on examining these studies cumulatively, is that intact ricin IT enter via their specific receptor. Lactose bound to the B-chain could be released either due to pH change in the endosome or to competition for the galactose binding site by galactose residues present in the Golgi. Ricin would then interact with the Golgi membrane, permitting insertion and translocation from this organelle. Since RTA IT exhibit at least minimal toxicity, they must have a translocation mechanism of their own, independent of B-chain. Translocation of RTA IT may be potentiated by RTA containing two sizable hydrophobic domains which are able to penetrate the membrane.[40,119,120] B-chain of ricin might then enhance the toxicity of RTA primarily by interacting with the intracellular vesicles from which RTA translocates and preventing lysosomal transfer and degradation of RTA. The B-chain could thereby function to promote the translocation of A-chain at the Golgi level. In this model, RTA IT would initially be trafficked like intact ricin IT. However, due to the lack of the galactose binding site, RTA IT would not interact with the Golgi membrane. Less IT would be associated with the Golgi and more would be transmitted to lysosomes and degraded. If correct, this model would argue for the use of genetically or biochemically modified B-chain in the construction of improved IT.

ACKNOWLEDGMENTS

This work was supported in part by National Institutes of Health grants R01-CA-31618, R01-CA-36725, and P01-CA-21737, American Cancer Society Research Grant IM-502, and the Minnesota Medical Foundation; it is Center for Experimental Transplantation and Cancer Research publication #46. D. A. Vallera is a Leukemia Society of America Scholar. The editorial skills of M. J. Hildreth are appreciated.

REFERENCES

1. **Vallera, D. A.,** Immunotoxins for ex vivo bone marrow purging in human bone marrow transplantation, in *Immunotoxins,* Frankel, A. E., Ed., Martinus Nijhoff, Boston, 1988, 515.
2. **Youle, R. J. and Neville, D. M., Jr.,** Kinetics of protein synthesis in activation by ricin-anti-Thy-1.1 monoclonal antibody hybrids: role of the ricin B subunit demonstrated by reconstitution, *J. Biol. Chem.,* 257, 1598, 1982.
3. **Vallera, D. A., Quinones, R. R., Azemove, S. M., and Soderling, C. C. B.,** Monoclonal antibody-toxin conjugates reactive against human T lymphocytes: a comparison of antibody linked to intact ricin toxin to antibody linked to ricin A chain, *Transplantation,* 37, 387, 1984.
4. **Royston, I., Madja, J. A., Baird, S. M., Meserve, B. L., and Griffiths, J. C.,** Monoclonal antibody specific for human T lymphocytes: identification of normal and malignant T cells, *Blood,* 54 (Suppl. 1), 106a, 1979.

5. **Weil-Hillman, G., Runge, W., Jansen, F. K., and Vallera, D. A.,** Cytotoxic effect of anti-M_r 67,000 protein immunotoxins on human tumors in a nude mouse model, *Cancer Res.,* 45, 1328, 1985.
6. **Vallera, D. A. and Myers, D. E.,** Immunotoxins containing ricin, in *Immunotoxins,* Frankel, A. E., Ed., Martinus Nijhoff, Boston, 1988, 141.
7. **Stong, R. C., Uckun, F. M., Youle, R. J., Kersey, J. H., and Vallera, D. A.,** Use of multiple T-cell directed intact ricin immunotoxins for autologous bone marrow transplantation, *Blood,* 66, 627, 1985.
8. **Quinones, R. R., Youle, R. J., Kersey, J. H., Zanjani, E. D., Azemove, S. M., Soderling, C. C. B., LeBien, T. W., Beverley, P. C. L., Neville, D. M., Jr., and Vallera, D. A.,** Anti-T cell monoclonal antibodies conjugated to ricin as potential reagents for human GVHD prophylaxis: effect on the generation of cytotoxic T cells in both peripheral blood and bone marrow, *J. Immunol.,* 132, 678, 1984.
9. **Stong, R. C., Youle, R. J., and Vallera, D. A.,** Elimination of clonogenic T-leukemic cells from human bone marrow using anti-M_r 65,000 protein immunotoxins, *Cancer Res.* 44, 3000, 1984.
10. **Vallera, D. A., Ash, R. C., Zanjani, E. D., Kersey, J. H., LeBien, T. W., Beverly, P. C. L., Neville, D. M., Jr., and Youle, R. J.,** Anti-T-cell reagents for human bone marrow transplantation: ricin linked to three monoclonal antibodies, *Science,* 222, 512, 1983.
11. **Siena, S., Villa, S., Bonadonna, G., Bregni, M., and Gianni, A. M.,** Specific *ex-vivo* depletion of human bone marrow T-lymphocytes by an anti-pan-T-cell (CD5) ricin A-chain immunotoxin, *Transplantation,* 43, 421, 1987.
12. **Casellas, P., Canat, X., Fauser, A. A., Gros, O., Laurent, G., Poncelet, P., and Jansen, F. K.,** Optimal elimination of leukemia T cells from human bone marrow with T101-ricin A-chain immunotoxin, *Blood,* 65, 289, 1985.
13. **Gorin, N. C., Douay, L., Laporte, J. P., Lopez, M., Zittoum, R., Rio, B., Daid, R., Stachowiak, J., Jansen, F. K., Casellas, P., Poncelet, P., Piance, M. C., Vioson, G. A., Salmon, C., LeBlanc, G., Deloux, J., Najma, A., and Duhamel, G.,** Autologous bone marrow transplantation with marrow decontaminated by immunotoxin T101 in the treatment of leukemia and lymphoma: first clinical observations, *Cancer Treat. Rep.,* 69, 953, 1985.
14. **Laurent, G., Poncelet, P., Fauser, A., Cassellas, P., Gorin, N. C., and Jansen, F. K.,** *Ex-vivo* treatment of bone marrow with T101 A-chain immunotoxin in autologous and allogeneic bone marrow transplantation, *Exp. Hematol.,* 12, 424, 1984.
15. **Manske, J. M., Buchsbaum, D. J., Azemove, S. M., Hanna, D. E., and Vallera, D. A.,** Antigenic modulation by anti-CD5 immunotoxins, *J. Immunol.,* 136, 4721, 1986.
16. **Laurent, G., Pris, J., Farcet, J.-P., Carayon, P. Blythman, H., Cassallas, P., Poncelet, P., and Jansen, F. K.,** Effects of therapy with T101 ricin A-chain immunotoxin in two leukemia patients, *Blood,* 67, 1680, 1986.
17. **Manske, J. M., Buchsbaum, D. J., Hanna, D. E., and Vallera, D. A.,** Cytotoxic effects of anti-CD5 radioimmunotoxins on human tumors *in vitro* and in a nude mouse model, *Cancer Res.,* 48, 7107, 1988.
18. **Weil-Hillman, G., Uckun, F. M., Manske, J. M., and Vallera, D. A.,** Combined immunochemotherapy of human solid tumors in nude mice, *Cancer Res.,* 47, 579, 1987.
19. **Casellas, P., Bourrie, J. P., Gros, P., and Jansen, F. K.,** Kinetics of cytotoxicity induced by immunotoxins: enhancement by lysosomotropic amines and carboxylic ionophores, *J. Biol. Chem.,* 259, 9359, 1984.
20. **Carriere, D., Casellas, P., Richer, G., Gros, P., and Jansen, F. K.,** Endocytosis of an antibody-ricin A chain conjugate (immuno-A-toxin) adsorbed on colloid gold, *Exp. Cell Res.,* 156, 327, 1985.
21. **Olsnes, S. and Sandvig, K.,** Entry of polypeptide toxins into animal cells, in *Endocytosis,* Pastan, I. and Willingham, M., Eds., Plenum Press, New York, 1985, 195.
22. **Baenziger, J. U. and Fiete, D.,** Structural determinants of *Ricinus communis* agglutinin and toxin specificity for oligosaccharides, *J. Biol. Chem.,* 254, 9795, 1979.
23. **Sandvig, K., Olsnes, S., and Pihl, A.,** Kinetics of binding of the toxic lectins abrin and ricin to surface receptors on human cells, *J. Biol. Chem.,* 251, 3977, 1976.
24. **Olsnes, S., Fernandez-Puentes, C., Carrasco, L., and Vasquez, D.,** Ribosome inactivation by the toxic lectins abrin and ricin. Kinetics of the enzymic activity of the toxin A-chains, *Eur. J. Biochem.,* 60, 218, 1975.
25. **Eiklid, K., Olsnes, S., and Pihl, A.,** Entry of lethal doses of abrin, ricin, and modeccin into the cytosol of HeLa cells, *Exp. Cell Res.,* 126, 321, 1980.
26. **Endo, Y., Kazuhiro, M., Motizuki, M., and Tsurugi, K.,** The mechanisms of action of ricin and related toxic lectins on eukaryotic ribosomes. The site and characteristics of the modification in 28S ribosomal RNA caused by the toxins, *J. Biol. Chem.,* 262, 5908, 1987.
27. **Olsnes, S. and Pihl, A.,** Chimeric toxins, *Pharmacol. Ther.,* 15, 355, 1982.
28. **Lin, J.-Y., Kao, W.-Y., Tserng, K.-Y., Chen, C.-C., and Tung, T.-C.,** Effect of crystalline abrin on the biosynthesis of protein, RNA, and DNA in experimental tumors, *Cancer Res.,* 30, 2431, 1970.
29. **Lin, J.-Y., Liu, K., Chen, C.-C., and Tung, T.-C.,** Effect of crystalline ricin on the biosynthesis of protein, RNA, and DNA in experimental tumor cells, *Cancer Res.,* 31, 921, 1971.

30. **Refsnes, K., Haylett, T., Sandvig, K., and Olsnes, S.,** Modeccin—a plant toxin inhibiting protein synthesis, *Biochem. Biophys. Res. Commun., 79*, 1176, 1977.

31. **Olsnes, S., Sandvig, K., Refsnes, K., and Pihl, A.,** Rates of different steps involved in the inhibition of protein synthesis by the toxic lectins abrin and ricin, *J. Biol. Chem., 251*, 3985, 1976.

32. **Sandvig, K. and Olsnes, S.,** Effect of temperature on the uptake, excretion and degradation of abrin and ricin by HeLa cells, *Exp. Cell Res., 121*, 15, 1979.

33. **Youle, R. J. and Neville, D. M., Jr.,** Anti-Thy 1.2 monoclonal antibody linked to ricin is a potent cell type specific toxin, *Proc. Natl. Acad. Sci. U.S.A., 77*, 5486, 1980.

34. **Vallera, D. A. and Uckun, F. M.,** Immunoconjugates, in *Biological Response Modifiers and Cancer Therapy,* Chiao, J. W., Ed., Marcel Dekker, New York, 1988, 17.

35. **Vitetta, E. S., Fulton, R. J., May, R. D., Till, M., and Uhr, J. W.,** Redesigning nature's poisons to create anti-tumor reagents, *Science, 238*, 1098, 1987.

36. **Leonard, J. E., Wang, Q.-C., Kaplan, N. O., and Royston, I.,** Kinetics of protein synthesis inactivation in human T-lymphocytes by selective monoclonal antibody-ricin conjugates, *Cancer Res., 45*, 5263, 1985.

37. **McIntosh, D. P., Edwards, D. C., Cumber, A. J., Parnell, G. D., Dean, C. J., Ross, W. C., and Forrester, J. A.,** Ricin B chain converts a non-cytotoxic antibody-ricin A chain conjugate into a potent and specific cytotoxic agent, *FEBS Lett., 164*, 17, 1983.

38. **Theisen, H.-J., Juhl, H., and Arndt, R.,** Selective killing of human bladder cancer cells by combined treatment with A and B chain ricin antibody conjugates, *Cancer Res., 47*, 419, 1987.

39. **Vitetta, E. S., Cushley, W., and Uhr, J. W.,** Synergy of ricin A chain-containing immunotoxins and ricin B chain-containing immunotoxins *in vitro* killing of neoplastic human B cells, *Proc. Natl. Acad. Sci. U.S.A., 80*, 6332, 1983.

40. **Calafat, J., Molthoff, C., and Hilkens, J.,** Endocytosis and intracellular routing of an antibody-ricin A chain conjugate, *Cancer Res., 48*, 3822, 1988.

41. **Vitetta, E. S.,** Synergy between immunotoxins prepared with native ricin A chains and chemically-modified ricin B chains *J. Immunol., 136*, 1880, 1986.

42. **Youle, R. J., Murray, G. J., and Neville, D. M., Jr.,** Studies on the galactose-binding site of ricin and the hybrid toxin Man-6P-ricin, *Cell, 23*, 551, 1981.

43. **Leonard, J. E., Wang, Q. C., Kaplan, N. O., and Royston, I.,** Kinetics of protein synthesis in human T-lymphocytes by selective monoclonal-antibody ricin conjugates, *Cancer Res., 45*, 5263, 1985.

44. **Moor, G. E., Woods, L. K., Minowada, J., and Mitchen, J. R.,** Establishment of a leukemia cell line with T cell characteristics, *In Vitro, 8*, 434, 1973.

45. **Royston, I., Majda, J. A., Baird, S. M., Meserve, B. L., and Griffiths, J. C.,** Human T cell antigens defined by monoclonal antibodies: the 65,000-dalton antigen of T cells (T65) is also found on chronic lymphocytic cells bearing surface immunoglobulin, *J. Immunol., 125*, 725, 1980.

46. **Caligaris-Cappio, F., Gobbi, M., Bofill, M., and Janossy, G.,** Infrequent normal B cells express features of B-chronic lymphocytic leukemia, *J. Exp. Med., 155*, 623, 1982.

47. **Gobbi, M., Caligaris-Cappio, F., and Janossy, G.,** Normal equivalent cells of B cell malignancies: analysis with monoclonal antibodies, *Br. J. Hematol., 54*, 393, 1983.

48. **Antin, J. H., Emerson, S. G., Martin, P., Gadol, N., and Ault, K. A.,** Leu-1+ (CD5+) B cells. A major lymphoid subpopulation in human fetal spleen: phenotypic and functional studies, *J. Immunol., 136*, 505, 1986.

49. **Bergman, Y. and Levy, R.,** Biosynthesis and processing of a human T lymphocyte antigen, *J. Immunol., 128*, 1334, 1982.

50. **Bernard, A. and Boumsell, L.,** The clusters of differentiation defined by the International Workshop on Human Leukocyte Differentiation Antigens, *Hum. Immunol., 11*, 1, 1984.

51. **Ledbetter, J. A., Parsons, M., Martin, P. J., Spooner, C. E., Wofsy, D., Tsu, T. T., Beatty, P. G., and Gladstone, P.,** Antibodies to Tp67 and Tp44 augment and sustain proliferative responses of activated T cells, *J. Immunol., 137*, 3229, 1986.

52. **Ceuppens, J. L. and Baroja, M. L.,** Monoclonal antibodies to the CD5 antigen can provide the necessary second signal for activation of isolated resting T cells by solid-phase-bound OKT3, *J. Immunol., 137*, 1816, 1986.

53. **Filipovich, A. H., Vallera, D. A., Youle, R. J., Haake, R., Blazar, B. R., Neville, D. M., Jr., Ramsay, N. K. C., McGlave, P., and Kersey, J. H.,** Graft-versus-host disease prevention in allogeneic bone marrow transplantation from histocompatible siblings: a pilot study using immunotoxins for T cell depletion of donor bone marrow, *Transplantation, 44*, 62, 1987.

54. **Vallera, D. A., Kersey, J. H., Quinones, R. R., Zanjani, E. D., Soderling, C. C. B., Azemove, S. M., LeBien, T. W., Beverley, P. C. L., Ash, R. C., Neville, D. M., Jr., and Youle, R. J.,** Antibody-ricin conjugates: purgative reagents for murine and human allogeneic bone marrow transplantation, in *Recent Advances in Bone Marrow Transplantation.,* Gale, R. P., Ed., Alan R. Liss, New York, 1983, 209.

55. **Vallera, D. A.,** The use of immunotoxins in bone marrow transplantation: eradication of T cells and leukemic cells, in *Immunoconjugates. Antibody Conjugates in Radioimaging and Therapy of Cancer,* Vogel, C.-W., Ed., Oxford Unversity Press, New York, 1987, 217.

56. **Filipovich, A. H., Vallera, D. A., Youle, R. J., Quinones, R. R., Neville, D. M., Jr., and Kersey, J. H.,** *Ex vivo* treatment of donor bone marrow with anti-T cell immunotoxins for the prevention of graft-versus-host disease, *Lancet,* 8375, 469, 1984.

57. **Filipovich, A., Vallera, D., Youle, R., Neville, D., and Kersey, J.,** *Ex vivo* T-cell depletion with immunotoxins in allogeneic bone marrow transplantation—the pilot study for prevention of graft-versus-host disease, *Transplant Proc.,* 17, 442, 1985.

58. **Seon, B. K.,** Selective killing of human T leukemic cells by immunotoxins prepared with ricin A chain and monoclonal anti-human T cell leukemia antibodies, *Cancer Res.,* 44, 259, 1984.

59. **Kersey, J. H., Weisdorf, D., Nesbit, M. E., LeBien, T. W., Woods, W. G., McGlave, P. B., Kim, T., Vallera, D. A., Goldman, A. I., Bostrum, B., Hurd, D., and Ramsay, N. K. C.,** Comparison of autologous and allogeneic bone marrow transplantation for treatment of high-risk refractory acute lymphoblastic leukemia, *N. Engl. J. Med.,* 317, 461, 1987.

60. **Vitetta, E. S. and Uhr, J. W.,** Immunotoxins, *Annu. Rev. Immunol.,* 3, 197, 1985.

61. **Thomas, E. D.,** The role of marrow transplantation in the eradication of malignant disease, *Cancer,* 49, 1963, 1982.

62. **Filipovich, A. H., Ramsay, N. K. C., Hurd, D., Stong, R., Youle, R., Vallera, D. A., and Kersey, J. H.,** Autologous Bone Marrow Transplantation (BMT) for T Cell Leukemia and Lymphoma using Marrow Cleaning with Anti-T Cell Immunotoxins, Autologous Bone Marrow Transplantation Meeting, University degli Studi di Parma, Parma, Italy, 1985.

63. **Vallera, D. A., Youle, R. J., Neville, D. M., Jr., and Kersey, J. H.,** Bone marrow transplantation across major histocompatibility barriers. V. Protection of mice from lethal graft-versus-host disease by pretreatment of donor cells with monoclonal anti-Thy-1.2 coupled to the toxin ricin, *J. Exp. Med.,* 155, 949, 1982.

64. **Mitsuyasu, R., Champli, R., Ho, W., Winston, D., Feig, S., Wells, J., Terasaki, P., Billing, R., Weaver, M., and Gale, R. P.,** Prospective randomized controlled trial of ex vivo therapy of donor bone marrow with monoclonal antibody anti-T-cell antibody plus complement for prevention of graft-versus-host disease: a preliminary report, *Transplant. Proc.,* 17, 482, 1985.

65. **Krolick, K. A., Uhr, J. W., Slavin, S., and Vitteta, E. S.,** *In vivo* therapy of a murine B cell tumor (BCL₁) using antibody-ricin A chain immunotoxins, *Exp. Med.,* 155, 1797, 1982.

66. **Kishida, Y., Masuho, M., Saito, T., Hara, T., and Fuji, H.,** Ricin A-chain conjugated with monoclonal anti-L1210 antibody. In vitro and in vivo anti-tumor activity, *Cancer Immunol. Immunother.,* 16, 93, 1983.

67. **FitzGerald, D. J., Willingham, M. C., and Pastan, I.,** Antitumor effects of an immunotoxin made with Pseudomonas exotoxin in a nude mouse model of human ovarian cancer, *Proc. Natl. Acad. Sci. U.S.A.,* 83, 6627, 1986.

68. **Hwang, K. M., Foon, K. A., Cheung, P. H., Pearson, J. W., and Oldham, R. K.,** Selective anti-tumor effect on L10 hepatocarcinoma cells of a potent immunoconjugate composed of the A chain of abrin and a monoclonal antibody to a hepatoma-associated antigen, *Cancer Res.,* 44, 4578, 1984.

69. **Hara, H. and Seon, B.,** Complete suppression of *in vivo* growth of human leukemia cells by specific immunotoxins: nude mouse models, *Proc. Natl. Acad. Sci. U.S.A.,* 84, 3390, 1987.

70. **Pastan, I. and Willingham, M. C.,** Journey to the center of the cell: role of the receptosome, *Science,* 214, 504, 1981.

71. **Pastan, I. and Willingham, M. C.,** The pathway of endocytosis, in *Endocytosis,* Pastan, I. and Willingham, M. C., Eds., Plenum Press, New York, 1985, 1.

72. **Lewis, W. H.,** Pinocytosis, *Bull. Johns Hopkins Hosp.,* 49, 17, 1931.

73. **Willingham, M. C. and Yamada, S. S.,** A mechanism for the destruction of pinosomes in cultured fibroblasts: Piranhalysis, *J. Cell Biol.,* 78, 480, 1978.

74. **Goldstein, J. L. and Brown, M. S.,** Binding and degradation of low density lipoproteins by cultured human fibroblasts: comparison of cells from a normal subject and from a patient with homozygous familial hypercholesterolemia, *J. Cell Biol.,* 249, 5153, 1974.

75. **Goldstein, J. L., Anderson, R. G. W., and Brown, M. S.,** Coated pits, coated vesicles, and receptor mediated endocytosis, *Nature,* 279, 679, 1979.

76. **Roth, T. F. and Porter, K. R.,** Yolk protein uptake in the oocyte of the mosquito *Aedes aegypti* L., *J. Cell Biol.,* 20, 313, 1964.

77. **Keen, J. H.,** The structure of clathrin-coated membranes: assembly and disassembly, in *Endocytosis,* Pastan, I. and Willingham, M. C., Eds., Plenum Press, New York, 1985, 85.

78. **Willingham, M. C. and Pastan, I. H.,** The transit of epidermal growth factor through coated pits of the Golgi system, *J. Cell Biol.,* 94, 207, 1982.

79. **Goldstein, J. L., Brown, M. S., Anderson, R. G. W., Russell, D. W., and Schneider, W. J.,** Receptor-mediated endocytosis: concepts emerging from the LDL receptor system, *Annu. Rev. Cell Biol.,* 1, 1, 1985.

80. **Marsh, M. and Helenius, A.,** Adsorptive endocytosis of Semliki Forest virus, *J. Mol. Biol.,* 142, 439, 1980.

81. **Willingham, M. C. and Pastan, I.,** The receptosome: an intermediate organelle of receptor mediated endocytosis in cultured fibroblasts, *Cell,* 21, 67, 1980.

82. **Tycko, B. and Maxfield, F. R.,** Rapid acidification of endocytic vesicles containing alpha$_2$-macroglobulin, *Cell,* 28, 643, 1982.

83. **Helenius, A., Mellman, I., Wall, D., and Hubbard, A.,** Endosomes, *Trends Biochem. Sci.,* 8, 245, 1983.

84. **Pastan I. and Willingham, M. C.,** Receptor-mediated endocytosis and the Golgi, *Trends Biochem. Sci.,* 8, 250, 1987.

85. **Dickson, R. B., Beguinot, L., Hanover, J. A., Richard, N. D., Willingham, M. C., and Pastan, I.,** Isolation and characterization of a highly enriched preparation of receptosomes (endosomes) from a human cell line, *Proc. Natl. Acad. Sci. U.S.A.,* 80, 5335, 1983.

86. **Farquhar, M. G.,** Multiple pathways of exocytosis, endocytosis, and membrane recycling: validation of a Golgi route, *Fed. Proc.,* 42, 2407, 1983.

87. **Nicolson, G.,** Ultrastructural analysis of toxin binding and entry into mammalian cells, *Nature,* 251, 628, 1974.

88. **Nicolson, G. L., Lacorbiere, M., and Eckhart, W.,** Qualitative and quantitative interactions of lectins with untreated and neuraminidase-treated normal, wild type, and temperature-sensitive polyoma-transformed fibroblasts, *Biochemistry,* 14, 172, 1975.

89. **Ray, B. and Wu, H. C.,** Chinese hamster ovary cell mutants defective in the internalization of ricin, *Mol. Cell Biol.,* 2, 535, 1982.

90. **Hyman, R., Lacorbiere, M., Staverek, S., and Nicolson, G.,** Derivation of lymphoma variants with reduced sensitivity to plant lectins, *J. Natl. Cancer Inst.,* 53, 963, 1974.

91. **Nicolson, G. L., Robbins, J. C., and Human, R.,** Cell surface receptors and their dynamics in toxin-treated malignant cells, *J. Supramol. Struct.,* 4, 15, 1976.

92. **Nicolson, G. L. and Poste, G.,** Mechanism of resistance to ricin toxin in selected mouse lymphoma cell lines, *J. Supramol. Struct.,* 8, 235, 1978.

93. **Gonatas, N. K., Stieber, A., Kim, S. U., Graham, D. I., and Avrameas, S.,** Internalization of neuronal plasma membrane ricin receptors into the Golgi apparatus, *Exp. Cell Res.,* 94, 426, 1975.

94. **Gonatas, N. K., Kim, S. U., Stieber, A., and Avrameas, S.,** Internalization of lectins in neuronal GERL, *J. Cell Biol.,* 73, 1, 1977.

95. **Gonatas, J., Stieber, A., Olsnes, S., and Gonatas, N. K.,** Pathways involved in fluid phase and adsorptive endocytosis in neuroblastoma, *J. Cell Biol.,* 87, 579, 1980.

96. **Sandvig, K. and Olsnes, S.,** Entry of toxic proteins abrin, modeccin, ricin, and diphtheria toxin into cells. II. Effect of pH, metabolic inhibitors and ionophores and evidence for penetration from endocytic vesicles, *J. Biol. Chem.,* 257, 7504, 1982.

97. **Sandvig, K., Olsnes, S., and Pihl, A.,** Binding, uptake and degradation of the toxic proteins abrin and ricin by toxin resistant cells, *Eur. J. Biochem.,* 82, 13, 1978.

98. **van Deurs, B., Tønnessen, T. I., Peterson, O. W., Sandvig, K., and Olsnes, S.,** Routing of internalized ricin and ricin conjugates to the Golgi complex, *J. Cell Biol.,* 102, 37, 1986.

99. **Yokoyama, M., Nishiyama, F., Kawai, N., and Hirano, N.,** The staining of Golgi membranes with *Ricinus communis* agglutin-horseradish peroxidase conjugate in mice tissue cells, *Exp. Cell Res.,* 125, 47, 1980.

100. **van Deurs, B., Peterson, O. W., Olsnes, S., and Sandvig, K.,** Delivery of internalized ricin from endosomes to cisternal Golgi elements is a discontinuous, temperature sensitive process, *Exp. Cell Res.,* 171, 137, 1987.

101. **Yoshitake, S., Watanabe, K., and Funatsu, G.,** Limited hydrolysis of ricin D with trypsin in the presence of sodium dodecyl sulfate, *Agric. Biol. Chem.,* 43, 2193, 1979.

102. **Manske, J. M., Buchsbaum, D. J., and Vallera, D. A.,** The role of ricin B chain in the intracellular trafficking of anti-CD5 immunotoxins, *J. Immunol.,* 142, 1755, 1989.

103. **Boyse, E. A., Old, L. J., and Luell, S.,** Antigenic properties of experimental leukemia. II. Immunologic studies *in vivo* with C57BL/6 radiation-induced leukemias, *J. Natl. Cancer Inst.,* 31, 987, 1963.

104. **Old, L. J., Stockert, E., Boyse, E. A., and Kim, J. H.,** Antigenic modulation. Loss of TL antigen from cells exposed to TL antibody. Study of the phenomenon *in vitro, J. Exp. Med.,* 127, 523, 1968.

105. **Chatenoud, L. and Bach, J.-F.,** Antigenic modulation—a major mechanism of antibody action, *Immunol. Today,* 5, 20, 1984.

106. **Schreiner, G. F. and Unanue, E. R.,** Membrane and cytoplasmic changes in B lymphocytes induced by ligand-surface immunoglobulin interaction, *Adv. Immunol.,* 24, 37, 1976.

107. **Stackpole, C. W., Jacobson, J. B., and Lardis, M. P.,** Two distinct types of capping of surface receptors on mouse lymphoid cells, *Nature,* 248, 232, 1974.

108. **Raso, V., Ritz, J., Basala, M., and Schlossman, S. F.,** Monoclonal antibody-ricin A chain conjugate selectively cytotoxic for cells bearing the common acute lymphoblastic leukemia antigen, *Cancer Res.,* 42, 457, 1982.

109. **Ukkonen, P., Lewis, V., Marsh, M., Helenius, A., and Mellman, I.,** Transport of macrophage Fc receptors and Fc receptor-bound ligands to lysosomes, *J. Exp. Med.,* 163, 952, 1986.
110. **Shawler, D. S., Miceli, M. C., Wormsley, S. B., Royston, I., and Dillman, R. O.,** Induction of in vitro and in vivo antigenic modulation by the anti-human T-cell monoclonal antibody T101, *Cancer Res.,* 44, 5921, 1984.
111. **Schroff, R. W., Farrell, M. M., Klein, R. A., Oldham, R. K., and Foon, K. A.,** T65 antigen modulation in a phase I monoclonal antibody trial with chronic lymphocytic leukemia patients, *J. Immunol.,* 133, 1641, 1984.
112. **Schroff, R. W., Klein, R. A., Farrell, M. M., and Stevenson, H. C.,** Enhancing effects of monocytes on modulation of a lymphocyte membrane antigen, *J. Immunol.,* 133, 2270, 1984.
113. **Youle, R. J. and Colombatti, M.,** Hybridoma cells containing intracellular anti-ricin antibodies show ricin meets secretory antibody before entering the cytosol, *J. Biol. Chem.,* 262, 4676, 1987.
114. **Tartakoff, A. M.,** Perturbation of vesicular traffic with the carboxylic ionophore monensin, *Cell,* 32, 1026, 1983.
115. **Raso, V. and Lawrence, J.,** Carboxylic ionophores enhance the toxicity of ligand and antibody delivered ricin A-chains, *J. Exp. Med.,* 160, 1234, 1984.
116. **Basu, S. K., Goldstein, J. L., Anderson, R. G. W., and Brown, M. S.,** Monensin interrupts the recycling of low density lipoprotein receptors in human fibroblasts, *Cell,* 24, 493, 1981.
117. **Raso, V., Watkins, S. C., Slayter, H., and Fehrmann, C.,** Intracellular pathways of ricin A chain cytotoxins, *Ann. N. Y. Acad. Sci.,* 507, 172, 1987.
118. **Press, O. W., Vitetta, E. S., Farr, A. G., Hansen, J. A., and Martin, P. J.,** Evaluation of ricin A-chain immunotoxins directed against human T cells, *Cell. Immunol.,* 102, 10, 1986.
119. **Uchida, T., Mekada, E., and Okada, Y.,** Hybrid toxin of the A chain of ricin toxin and a subunit of *Wistaria floribunda* lectin. Possible importance of the hydrophobic region for entry of toxin into the cell, *J. Biol. Chem.,* 255, 6687, 1980.
120. **Ishida, B., Cawley, D. B., Reue, K., and Wisnieski, B. J.,** Lipid-protein interactions during ricin toxin insertion into membranes. Evidence for A and B chain penetration, *J. Biol. Chem.,* 258, 5933, 1983.

Chapter 9

TOXIN CONJUGATES: INTERACTIONS WITH MAMMALIAN CELLS

David FitzGerald and Ira Pastan

TABLE OF CONTENTS

I. DEFINITION

A toxin conjugate is a hybrid molecule where one part of the hybrid is derived from a protein toxin and the other from an unrelated second protein, which sometimes will be referred to as the "carrier protein".

II. INTRODUCTION

Protein toxins are potent biological effector agents. While many toxins can be and are being used to make conjugates, this review focuses mainly on the bacterial toxins pseudomonas exotoxin (PE) and diphtheria toxin (DT) and the plant toxin ricin. These three toxins bind to cells, enter by receptor-mediated encocytosis (RME), and then act catalytically in the cell cytoplasm to inhibit protein synthesis.[1,2]

To initiate their interaction with mammalian cells, toxins first bind to a component of the cell surface. Toxin receptors that mediate this interaction have been identified for some toxins, e.g., the binding of cholera toxin to GM1 ganglioside. However, in most cases, the existence of specific receptors has only been inferred from binding studies using radiolabeled toxin. Once bound, toxins enter the cell by endocytosis and must reach the cytoplasm by membrane translocation. After being delivered to their site of action within the cytoplasm, the toxins can exert a biochemical effect on the cell. PE and DT shut down protein synthesis by ADP-ribosylating elongation factor two (EF-2). Ricin has N-glycosidase activity which removes a single adenine residue from ribosomal RNA.[3] Thus, we can expect toxins to have at least three functional domains: receptor binding, membrane translocation, and enzyme activity.

Traditionally, toxin conjugates have been made by chemically cross-linking a toxin with an unrelated second protein. More recently, it has been possible to make conjugates by gene-fusion techniques, whereby the DNA coding for a portion of the toxin is fused to the carrier protein DNA or cDNA. Toxin conjugates can be made either with a whole toxin molecule containing all three functional domains, or with a toxin subunit containing a subset of these functions.

Toxin conjugates have some practical applications. Conjugates are used frequently to learn more about the nature of the toxin molecules themselves. For instance, the binding domain of a given toxin can be replaced by a protein or peptide with a different binding specificity than the toxin. By comparing the cell-killing activities of the native toxin with those of the conjugate, the contribution made by the binding domain of the native toxin can be assessed. Likewise, we can learn more about the biology of the carrier protein. For instance, if a toxin-antibody conjugate has low cell-killing activity, this might indicate that the antibody is poorly internalized.

Besides using these conjugates to gain insight into various aspects of cell biology, there is interest in developing toxin-based proteins as novel drugs for clinical medicine. Antibody-toxin conjugates, called immunotoxins, may be useful in treating malignancies, graft vs. host disease, autoimmune states, and possibly even parasitic diseases (for reviews of immunotoxins, see References 4 to 8). Hormone, growth factor, and interleukin conjugates with toxins may be used to eliminate populations of cells expressing high levels of these receptors; and antigen-toxin conjugates could be useful in inducing immune tolerance to specific antigens.[9]

There are some potential problems with using toxin conjugates in clinical settings. One is whether the immunogenicity of the toxin portion will prevent repeated dosings; another is whether the large size of some conjugates will prohibit access to target cells. Finally, there is a general concern about nonspecific toxicity mediated by the toxin portion of the conjugates. Since the administration of toxin-based reagents in clinical settings is still in its infancy, it is not clear whether these concerns will impede general use or not.

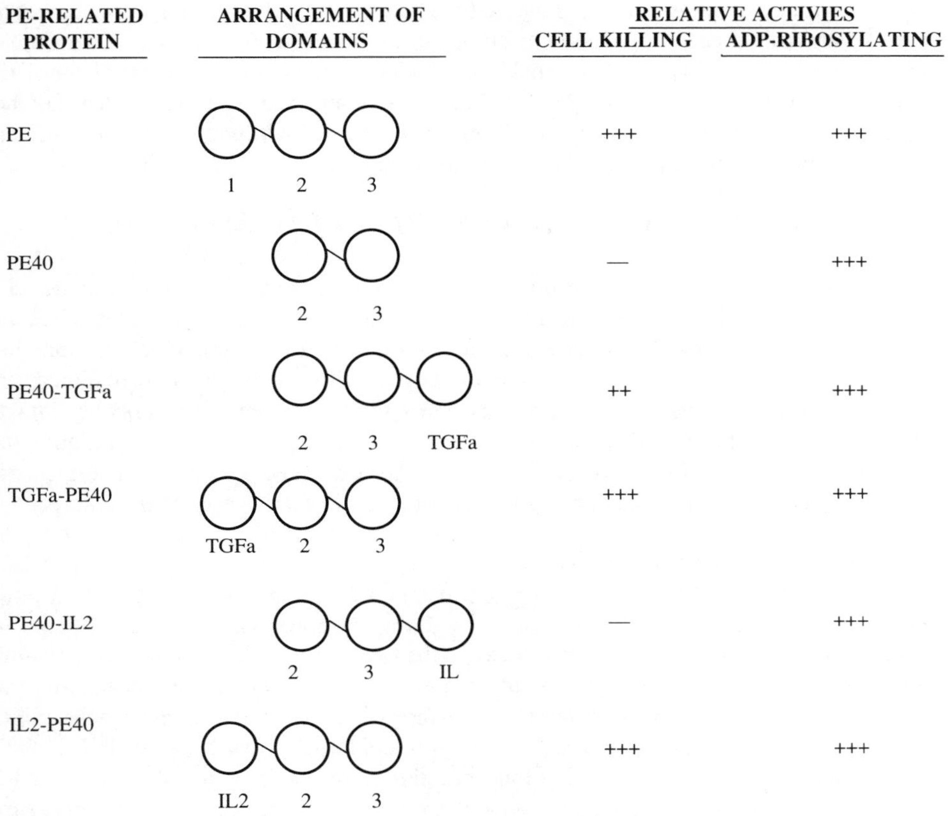

PE-RELATED PROTEIN	ARRANGEMENT OF DOMAINS	RELATIVE ACTIVIES	
		CELL KILLING	ADP-RIBOSYLATING
PE	1 2 3	+++	+++
PE40	2 3	—	+++
PE40-TGFa	2 3 TGFa	++	+++
TGFa-PE40	TGFa 2 3	+++	+++
PE40-IL2	2 3 IL	—	+++
IL2-PE40	IL2 2 3	+++	+++

FIGURE 1. Activities of PE-related proteins. Relative cell-killing and ADP-ribosylating activities of various PE-related proteins were determined. Cell-killing activity of: PE and PE40 was assayed on Swiss 3T3 cells; PE40-TGFα and TGFα-PR40 on A431 cells; and PE40-IL2 and IL2-PE40 on HUT-102 cells. ADP-ribosylating activity was measured as described in Reference 24.

III. TOXIN STRUCTURE

By the nature of the endeavor, only those toxins which have been purified and reasonably well characterized have been used to make toxin conjugates; toxins that inhibit protein synthesis have been used most frequently. These include a class of plant proteins termed ribosome-inactivating proteins, such as ricin, abrin, gelonin, and pokeweed antiviral protein and the bacterial toxins PE and DT.

The prototype toxin has an A-chain (A for activity) and a B-chain (B for binding to cells). Gelonin and pokeweed antiviral proteins are exceptions and occur as natural A-chain toxins. While the traditional A- and B-chain model is still useful at times, it does not do justice to the complex nature of these toxin molecules. Recently, X-ray diffraction analysis of PE and ricin toxin crystals has demonstrated the presence of three and five structural domains, respectively.[10,11]

We and others have assigned specific functions to the three domains of PE.[12-15] Reading from the amino to the carboxyl end, domain I binds to cells, domain II is needed for membrane translocation, and domain III contains ADP-ribosylating activity (see Figure 1). The crystal structure of DT has not been reported yet, but functionally DT has the same three domains as PE but arranged in the opposite orientation with respect to the amino and carboxyl termini.[2]

Like PE, ricin protein has been crystallized and its structure analyzed. Three structural domains are found in the A-chain and two in the B-chain.[11] The two B-chain domains appear to have arisen from a gene duplication; each one can bind a galactoside. One of the domains of the A-chain presumably has *N*-glycosidase activity. While we know the A-chain of ricin has the enzyme activity and the B-chain binds, the translocation function has not been assigned wholly to either chain. In fact, both chains have hydrophobic residues and each may need to be present to achieve the full translocating activity of native ricin.[4]

IV. CONSTRUCTION OF TOXIN CONJUGATES

The first toxin conjugates were probably molecular aggregates. Homobifunctional reagents such as glutaldehyde were used to cross-link toxins with carrier proteins of interest. The construction of discrete conjugates became possible with the advent of heterobifunctional reagents. Ultimately, however, heterobifunctional cross-liners may not be ideal either, since the chemical modification of some proteins alters their behavior when injected into test animals. To avoid these problems, new approaches using gene-cloning techniques have produced "molecular fusions" where the two proteins are linked by virtue of being encoded by a single piece of DNA. This latter approach appears to be the way of the future.

A. CHEMICAL CROSS-LINKING

Toxins have been chemically conjugated to monoclonal antibodies and to a variety of growth factors, hormones, and serum proteins. Using carefully controlled conditions, it is possible to make one-to-one conjugates between the toxin and carrier protein. Primary amino groups and sulfhydryl groups are most often used as the reactive groups in making these conjugates. Most proteins have at least one lysine which can serve as the reactive amino group, but if there is no lysine in the protein, as in epidermal growth factor (EGF), the N-terminal amino group can be used.[16] Many proteins have disulfide bonds, but fewer have free sulfhydryl groups. Therefore, a frequent strategy is to introduce a sulfhydryl group using specialized reagents, such as *N*-succinimidyl-3-(2-pyridyldithio)propionate (SPDP) or 2-iminothiolane.[8,7,18] Generally, the two proteins, toxin and carrier, are linked by either a disulfide bond or a thioether bond and either methodology can usually be used to make the desired conjugate. Disulfide conjugates can be made by using either SPDP as the heterobifunctional cross-linker or by using a combination of 2-iminothiolane followed by Ellman's reagent. To make a thioether type of linkage, a different kind of cross-linker is needed. Typically, the cross-linker will have both a reactive succinimide ester and reactive maleimide ester. MBS is a popular choice for such a reagent.[19]

The choice of whether to use disulfide or thioether linkage is usually based on functional assays. For instance, the A-chain of ricin (RTA) can be linked to various monoclonal antibodies by either a disulfide or thioether linkage. However, only the disulfide conjugates are active in cell-killing assays.[4] While thioether conjugates with A-chain produce conjugates that have poor cell-killing activity, similar conjugates with whole ricin are usually very potent. It is not well understood why RTA immunotoxins have poor cell-killing activity when a thioether linkage is used.

DT, PE, or their subunits can be cross-linked to carriers by thioether or disulfide linkage without apparent loss of cell-killing activity.[19,20,21,22] The fact that either linkage type can result in an active conjugate raises the question of how the ADP-ribosylating portion of these toxins is freed from its carrier protein and delivered to the cell cytoplasm. Disulfide-linked proteins are thought to separate into their component parts by, as yet poorly defined, reduction systems, but cleavage of a thioether bond has not been documented. If the thioether bond is not broken, possibly a portion of the toxin is separated from the rest of the conjugate by a proteolytic clip within the toxin.

It is not clear yet which type of linkage will be best suited for use in clinical settings.

TABLE 1
Cytotoxic Activity of Various PE-Related Conjugates Directed against the EGF Receptor
(ID_{50},[a] ng/ml, 24 h)

Cell line	EGF-PE	PE40-TGFα	TGFα-PE40
A431	0.4	4.0	0.4

[a] Concentration of toxin-related material that inhibits by 50%.

Some reports have suggested that disulfide-linked proteins, made using the SPDP/2-iminothiolane methodology, may not be stable *in vivo*. To overcome this problem, Thorpe et al.[23] have designed a cross-linker that generates a hindered disulfide bond. Using such an approach they have demonstrated prolonged *in vivo* half-lives for RTA immunotoxins. Acid-labile cross-linkers also have been described but have not been used extensively. In this case, the reagent, which is based on 2-methylmaleic anhydride, forms an acid-labile linker upon reaction with amino groups.[24]

B. CONJUGATES MADE BY GENE FUSION

Using recombinant DNA technology, the construction of gene fusions has become commonplace. For toxin work, however, it has been confined to those few toxins that have been cloned, and specific examples of toxin gene fusions have been documented only recently. The first description of a toxin-related gene fusion was the replacement of the binding portion (the carboxyl end) of DT with the structural gene coding for alpha melanocyte-stimulating hormone (MSH).[25] The goal was to design a toxin conjugate that would bind to cells via the MSH receptor and selectively kill receptor-positive cells. Murphy et al.[25] showed that such a chimeric protein could kill receptor-positive melanoma (NEL-M1) cells but not receptor-negative CV-1 or CHO cells.

Since the initial report with alpha MSH, additional toxin-related gene fusions have been made. cDNAs for either transforming growth factor (TGF)α or interleukin 2 were fused to DNA coding for a portion of the PE structural gene containing domains II and III (termed PE40) and lacking the binding domain (see Figure 1).[26,27] We had shown previously that it was possible to link PE chemically to EGF and produce a conjugate that specifically killed cells with high numbers of EGF receptors. We made this conjugate by reacting the cross-linker with the N-terminal amino group of EGF. Because of this we reasoned that the amino terminus might not be involved directly in cell recognition. Therefore, initially, the cDNA for TGFα was placed 3' to the coding region of PE40 (see Figure 1).[26]

To make sufficient protein for analysis, DNA encoding this fusion protein, PE40-TGFα, was cloned into a T_7 promoter-driven bacterial expression system. PE40-TGFα fusion protein was prepared and assayed for cytotoxic activity on EGF receptor-positive and receptor-negative cell lines. Its activity was compared with the chemical conjugate, PE-EGF. The gene fusion product had one tenth the cell-killing activity of PE-EGF but showed very little toxicity on receptor-negative cell lines.

The gene order of the fusion was later changed so that TGFα was placed 5' relative to PE40.[15] Thus, in the fusion protein, TGFα would have the same location relative to domains II and III as domain I has (see Figure 1). TGFα-PE40 was purified and the activity compared with PE40-TGFα and PE-EGF. The result of this comparison indicated that TGFα is in a more favorable location when placed on the amino end of the fusion protein. TGFα-PE40 was as active as PE-EGF and tenfold more active than PE40-TGFα (Table 1). Thus, the positioning of the binding domain may be important for cell-killing activity of PE-related conjugates.

Similar constructions were made with PE40 and a cDNA for IL2.[27] As with TGFα, IL2 coding sequences were placed either 5' or 3' to the PE40 sequence. The importance of relative position for each component was even more dramatic than with TGFα. When IL2 was placed 3' to PE40, there was no cell-killing activity when the fusion protein was tested on IL2 receptor-positive cells. However, expressing the fusion protein with IL2 at the amino end produced a potent cell-killing protein with ID_{50} of 20 ng/ml for various IL2-receptor-positive cell lines (see Figure 1). To date, no chemical conjugates between PE and IL2 have been made, so a comparison between PE-IL2 and IL2-PE40 has not been performed.

Recently, a recombinant human CD4-PE40 hybrid protein has been used to selectively kill HIV-infected cells *in vitro*.[27a] In this hybrid molecule, the cell recognition domain (1) of PE is replaced with part of CD4. CD4 is the T-lymphocyte surface molecule which serves as the HIV receptor. This hybrid protein was synthesized in *Escherichia coli,* purified, and shown to react with polyclonal antibodies against both CD4 and PE; it was also shown to have ADP-ribosylation activity. CD4-PE40 was able to selectively kill cells expressing the HIV envelope glycoprotein. These results suggest that such hybrid proteins may represent novel therapeutic agents to be used in the treatment of acquired immune deficiency syndrome.

Gene fusion technology for production of toxin conjugates is relatively new, but it appears to be an approach with certain advantages over chemical linkage: (1) a homogenous product is made; (2) there are no unreacted cross-linker groups; (3) unwanted protein sequences that may lead to nonspecific toxic effects can be removed at the DNA level; and (4) large-scale production is possible in a suitable expression system.

V. TOXIN FUNCTION

When the entire binding portion of a toxin is removed, inactivated, or blocked, cell-killing activity can be reduced by as much as 10,000-fold.[19,28] Thus, toxins rely heavily on cell binding to achieve high potency. Replacing the native toxin binding domain with a novel binding protein can restore cell-killing capability, but the sensitivity of target cells to the conjugate then reflects the binding affinity of the new protein. By comparing the cell-killing activities of the native toxin with those of the conjugate, it is possible to analyze the role of the toxin binding component in cell killing. However, while the notion of freely substituting one binding component for another can be informative, there are some situations where such simple substitutions may not reflect the complex interaction between toxins and cells.

DT appears to have three functional regions.[1,28] The amino portion of this toxin, the A-chain, is a 21,000-mol wt protein that has ADP-ribosylating activity. The remainder of the molecule has been termed the B-chain. The B-chain can be further divided into two parts: the binding portion of DT has been assigned to the carboxyl end of the B-chain, while the amino portion (of the B-chain) is rich in hydrophobic amino acids and may play a central role in membrane translocation to the cell cytoplasm. When the 17,000-mol wt carboxyl fragment of DT B-chain is removed, there is little residual toxicity for cells. Cross-reacting material CRM45 and MspSA are examples of such proteins that lack this 17,000-mol wt fragment and, along with A-chain and whole DT, have been conjugated to monoclonal antibodies as a way to study DT function.[19]

Youle, Greenfield, and colleagues have investigated the roles of the various domains of DT by making immunotoxins with either A-chain alone (DTA), A-chain plus the hydrophobic residues contained in CRM45 and MspSA, or with the entire toxin.[19] They chemically coupled these various forms of DT to the monoclonal antibody UCHT1, which recognizes the human T cell antigen T3. The results showed that the UCHT1-DT was 100-fold more active than UCHT1-MspSA and 10,000-fold more active than UCHT1-DTA. Such results indicate that the information contained in the hydrophobic residues that make up the carboxyl end of MspSA improve the efficiency of getting the A-chain to the cell cytoplasm. However, for full cell-killing activity, the entire DT molecule is needed. The interpretation of these results is not so simple. If the hydrophobic piece of the B-chain contained all the information for translocation, presumably there would be no difference in cell-killing activity between UCHT1-MspSA and UCHT1-DT. The 100-fold difference could be explained if the hydrophobic portion of the B-chain contained only part of the translocating domain and the other part was found in the carboxyl 17,000-mol wt portion. Alternatively, there might be residual DT binding activity when the conjugate is made with native DT and this might contribute to its greater potency. Finally, the carboxyl end of DT might not have an active role in translocation to the cytoplasm but might keep the protein conformation in the most favorable position to allow the hydrophobic domain to perform its function.

A second set of experiments by the same group examined point mutations in the carboxyl end of DT.[20] These point mutations were shown previously to reduce the cell-killing activity of native DT by mechanisms that were unknown at the time of their isolation.[29] CRM102, 103, and 107 were examined for cytotoxic and binding activity. The results indicated that CRM102 and 103 were 1/100 as toxic for Vero cells and Jurkat cells as native DT, while CRM107 was 1/10,000 as toxic. This loss of activity correlated with reduction in binding affinity by 1/100 for CRM102 and 103 and by 1/8000 for CRM107. These mutant forms of DT, as well as native DT, were chemically linked to UCHT1 and further analyzed on target and nontarget cells. CRM103, 107, and DT were all equally toxic for Jurkat cells when linked to UCHT1. However, when assayed on Vero cells, which have many DT receptors and are highly sensitive to DT, the UCHT1-DT conjugate gave 90% inhibition of protein synthesis at $6 \times 10^{-10} M$. UCHT1-CRM107 had at most 1/2000 of that activity on Vero cells. These results indicate that native DT conjugates retain some binding activity for DT receptors even after conjugation to an antibody. The non-antibody-mediated toxicity of UCHT1-DT conjugate was not present when CRM mutants were used that were deficient in binding activity.

Even in the absence of DT-mediated binding, a full length DT protein gives greater potency to the conjugate than MspSA or CRM45. This suggests that DT CRM proteins, containing all three domains but having a point mutation in the binding portion of the toxin might be best for making DT conjugates of high potency and low nonspecific activity.

The A- and B-chains of ricin are held together by a disulfide bond. The A-chain (RTA) is enzymatically active and can inhibit protein synthesis in a cell-free system, while the B-chain binds to cells by interacting with cell-surface terminal galactose residues. Binding can be inhibited by adding excess lactose. Ricin-related conjugates can be made using either the A-chain by itself or the native ricin protein.[30] When whole ricin is used, some provision must be made to block the nonspecific binding to cell-surface galactose residues.[30]

Ricin-related immunotoxins are frequently made by conjugating purified RTA or recombinant RTA to monoclonal antibodies.[6,31] RTA conjugates made with growth factors and serum proteins, such as EGF and transferrin also have been described.[32,33] RTA has both the translocating and the enzyme activity necessary to kill cells, but not the binding activity, so binding must be supplied by the carrier protein. The binding of monoclonal antibodies to receptors usually leads to internalization and when these antibodies are coupled with RTA, the resulting conjugate is usually a potent cell-killing agent.[34,35]

Some RTA immunotoxins bind cells but have no cell-killing activity. The simplest explanation for this would be that the antibody failed to induce internalization. However, a variety of data suggest that the situation is more complex. Enhancing agents (low molecular weight compounds that change the intracellular environments or destinations in a favorable way for toxins) can render a nontoxic RTA immunotoxin fully active even though there are no data to suggest that the mechanism of action of enhancers is to induce internalization.[36,37] Two features of RTA conjugates have been revealed by these studies. The first might seem self-evident but in fact is important to establish that RTA must be internalized to be active as a cell-killing agent. The second feature suggests that certain kinds of intracellular environments must be present, and that RTA must be delivered to these environments for the conjugate to be active. Unlike DT or PE, however, RTA does not appear to require an intracellular acidic environment for activity.

As mentioned above, it is also possible to make conjugates with "whole" ricin. These have been made in two ways: (1) conjugation which does not alter the galactose binding site and (2) conjugation which sterically blocks the galactose binding site. In experiments with the first type of conjugate, 100 mM lactose is added to the cell culture medium.[30] This prevents the B-chain from interacting with cell surface galactose residues. This approach cannot be used *in vivo*. Therefore, whole ricin conjugates whose galactose binding has been blocked by steric hindrance due to reaction of cross-linkers close to the sugar binding site also have been prepared.[38] These might be useful for systemic administration but have failed to gain general acceptance because of the fear of toxicity mediated by residual B-chain binding. The advantage of using whole ricin comes from the high cell-killing activity that is achieved when the toxin is coupled to most cell-binding monoclonal antibodies.[39]

Various kinds of conjugates have been made with PE including growth factor and antibody conjugates, conjugates made by chemical attachment using either disulfide or thioether linkage and conjugates made by gene fusions. As mentioned above, X-ray diffraction analysis of PE crystals has revealed three major structural domains. Expression of these domains in *E. coli* either singly or in combination has elucidated some of their functions. Confirmation of particular functions has recently been possible by making conjugates with various portions of PE.

The evidence that domain I (see Figure 1) of PE is responsible for binding has come from several experimental approaches. When cells were pretreated with purified domain I, the cell-killing activity of native PE was blocked.[12] When PE40 (which contains all of domains II and III and none of domain I) was added to cells, there was no cell killing even though full ADP-ribosylation activity was present (Figure 1). When cell binding was restored to PE40 by either chemical conjugation or gene fusion with another cell-binding protein, cell killing was restored.[21,26,27] The latter finding also implies that PE does not require interaction with PE receptors to mediate cell killing.

Domain III of PE contains the ADP-ribosylating activity of PE.[12,13,40] Expression of this domain by itself has confirmed this conclusion. To date, no conjugate has been made with domain III alone. Because it lacks translocating activity, the prediction would be that it would have poor cell-killing activity compared with a conjugate made with PE40. This hypothesis is currently being tested.

The role of domain II of PE has been difficult to determine. The expression of the plasmid pJH13, which contains a deletion in domain II, produced a protein with binding and ADP-ribosylating activity but little or no cell-killing activity.[12] A simple explanation would be that domain II was necessary for efficient translocation to the cell cytoplasm and that the deletion removed part of the required domain. However, it is also possible that deleting part of domain II indirectly caused the poor translocating activity. By making small deletions in domain II and further characterizing various mutant proteins by site-directed mutagenesis, it may be possible to determine the location of the residues needed for translocation. If domain II has "pure" translocating activity, it may be possible to use domain II to mediate the translocation of other proteins, including metabolic enzymes, into the cell cytoplasm. This kind of toxin conjugate could, in theory, be used for protein therapy to restore normal metabolic states to cells with inborn errors of metabolism.

VI. CARRIER PROTEIN FUNCTION

Antibody-toxin conjugates, immunotoxins, have been reviewed extensively in several places recently and so do not get an exhaustive treatment here.[4-8] Routinely, monoclonal antibodies (MoAbs) are selected to bind certain tissue types. Large amounts of antibody are purified and subsequently conjugated with the toxin of choice. The resulting immunotoxin is then assayed for cell-killing activity, usually on target and nontarget cells.

The fate of the antibody after binding to the target cell, while having a profound influence on the outcome of the cell-killing assay, unfortunately is poorly understood. Since toxins act intracellularly, antibodies must carry toxins into the cell if they are to be active cell-killing reagents. Two important questions remain to be addressed. (1) Can one assume that all active immunotoxins rely solely on antibody-mediated internalization for cell-killing or do toxins also play a role in cell entry? (2) Is the failure of an immunotoxin to kill cells due solely to a lack of antibody-mediated internalization or do intracellular conditions also dictate the ability of the toxin to reach the cytoplasmic target?

In the past 5 years, RTA has been coupled to a variety of MoAbs. A large percentage of these immunotoxins killed cells, but many did not. If the particular antibody was efficiently internalized, as most monoclonal antibodies to receptors are, then the RTA immunotoxin was usually quite active.[34,35] However, a number of the MoAbs were selected because they bound selectively to breast, ovarian, or colon cancer tissue. When these MoAbs were conjugated to RTA, some of these immunotoxins were found to be quite active and others were not.[41] The reasons for the disparate results are still not entirely clear. A straight-forward explanation would be that the active immunotoxins relied on antibody-mediated internalization and the nonactive ones failed to kill because the antibody remained immobile at the cell surface.[41] Thus, one can imagine at least three possible outcomes following the binding of an antibody: (1) the antibody could bind a receptor or other component that cycled into the cell continuously; (2) the antibody could trigger internalization by the very act of binding; and (3) antibody binding could fail to trigger internalization. Existing data suggest that inactive immunotoxins fail to kill cells at least in part because they are not internalized.

However, RTA, like other toxins, needs to reach a specific intracellular environment that affords the right conditions for getting the toxin from a vesicular structure into the cell cytoplasm. Experimental data also suggest that the specific intracellular destination is important for toxin activity. Enhancers can convert a nontoxic immunotoxin to one that is fully cytotoxic. If the enhancer promoted internalization, then cell killing could be explained at the level of internalization. However, there is no good evidence that enhancers promote translocation. Clearly, further study is needed.

Immunotoxins made with native PE have been shown to be 5- to 500-fold more active in cell-killing assays than corresponding RTA immunotoxins made with the same antibody.[41,42] Also, three monoclonal antibodies have been conjugated with PE40, a recombinant form of PE containing domains II and III but none of the binding domain.[21] Anti-Tac and HB21 are both monoclonal antibodies that bind cell surface receptors and are thought to be internalized following binding. When PE40 was conjugated to these antibodies, active cell-killing immunotoxins were produced. OVB3 binds to an uncharacterized antigen associated with ovarian cancer cells.[43] When it was coupled to native PE, a very active immunotoxin was made.[43] However, in the same assay, the immunotoxin made with OVB3 and PE40 was completely inactive.[21] This result with PE40 and PE and similar results comparing activities of RTA and whole ricin immunotoxins point toward native toxins being able to convert noninternalizing antibodies into active immunotoxins by the presence of a binding domain associated with the toxin. A complex aspect of these types of results is the fact that the addition of excess antibody can block the cell-killing activity of immunotoxins made with whole toxins even when there is residual binding contributed by the toxin.

Interpreting the results of cell-killing experiments which use antibodies conjugated to native toxin molecules is not easy. When an immunotoxin is added to cells, antibody binding brings the toxin molecule very close to the cell membrane and thereby increases the local concentration of the toxin around the cell. Interactive events between toxin and membrane that are concentration dependent may happen only at a low frequency when the toxin is in solution but quite often when the effective concentration is raised in this way. Weak binding interactions between toxins and cells, which normally are of no biologic significance, may play a significant role in immunotoxin action.

Data that indicate whether a monoclonal antibody is internalized or not may suggest additional uses for the antibody. An antibody which binds target cells strongly and makes a poor immunotoxin because the antibody is not internalized, might make an ideal candidate for a conjugate between a high energy radionuclide and that antibody. Antibodies that recognize receptors and are efficiently internalized might be used to transport toxins or drugs or to down-regulate specific receptors and render cells less sensitive to particular hormones or growth factors.

VII. TARGETING OF TOXIN CONJUGATES

Cells express on their cell surface a large variety of different macromolecules, such as proteins, glycoproteins, glycolipids, gangliosides, etc. Functionally, some of these serve as receptors, some as structural elements, some as identity markers, and some have unidentified purposes. Normal cells may express one subset of these surface molecules while malignant cells may express a different yet overlapping subset; maturing cells express differentiation antigens and lymphocytes express antigen recognition proteins. In theory, it should be possible to target toxin conjugates to any of these various cell-surface components.

To date, most targeting of toxins has been carried out to kill: (1) normal or malignant cells expressing a particular receptor, e.g., the EGF receptor,[15,16,32] the IL2 receptor,[27,35,44] the transferrin receptor,[33,34,45-47] the asialoglycoprotein receptor,[48] and the mannose-6-phosphate receptor;[49] (2) malignant cells expressing an antigen associated with malignancy, e.g., CEA;[50] or (3) lymphocytes expressing a particular receptor, antigen, or type of antibody.[51]

Antireceptor targeting could be useful in a number of ways, including the selection of receptor-defective or receptor-negative mutants.[52,53] By administering antireceptor conjugates *in vivo*, it might be possible to kill selectively and, therefore, map receptor-positive cells. Pathologic damage might reveal occult populations of cells expressing a given receptor. Also, elevated levels of some receptors have clearly been shown to be associated with diseased states. The EGF receptor appears to be an oncogene, since high level expression of the receptor alone or in combination with exogenous EGF can cause transformation.[54] High levels of the IL2 receptor are associated with adult T cell leukemia,[35] bombesin receptors with small cell carcinoma of the lung,[55] and high levels of transferrin receptors with rapidly dividing cells.[33,34,46] Toxin conjugates directed to these receptors may be useful to treat certain cancers.

Tumor-associated antigens that are specific for malignancy and expressed nowhere else or at no other time may not exist. However, it seems equally clear that various MoAbs bind some types of cancers and only a small number of normal tissues.[43] Such antibodies when coupled to toxins may serve as novel drugs for cancer therapy.

VIII. TOXIN CONJUGATES AS PHARMACEUTICALS

Since toxins kill most cell types, their value as conjugates will depend on efforts to redirect their killing activity to target cells and away from vital organs. Both RTA and PE40 appear to have minimal nonspecific toxicity and may be very useful agents with which to develop toxin-based drugs. Several DT-related polypeptides would also be suitable except that most individuals in the developed world have been vaccinated with diphtheria toxoid. This has resulted in a high percentage of individuals having circulating neutralizing antibodies to DT. However, DT-related toxins might be useful in veterinary medicine, in specific human populations that have never been vaccinated with diphtheria toxoid, or in individuals who have an undetectable humoral response to DT.

As of this writing, only ricin- and PE-derived toxins have been administered to humans. Other toxins will surely follow closely behind. Toxins, such as shigella toxin, barley toxin, and various clostridial toxins, may be suitable for conjugate development.

With some exceptions, the toxins and toxin subunits currently being considered for clinical use are polypeptides. They range in molecular weight from 30 to 66 kDa and when conjugated to IgG antibodies could form hybrid proteins with molecular weights as high as 210 kDa. Some investigators have expressed concern that molecules of such size might not be suitable for systemic administration because of possible problems in crossing capillary beds to gain access to target cells. Whether there is a size constraint affecting tissue permeability of conjugates, and what size that might be, has not yet been established. One solution to this potential problem is to begin administering conjugates in the form of regional therapy. Another strategy is to reduce the size of the conjugate by eliminating unwanted segments of either the toxin or carrier.

Regional therapy might be highly beneficial for patients with such diseases as ovarian cancer and brain malignancies. In those situations, the target cells usually remain confined to a body cavity and direct infusion of immunotoxins into this space might be optimal for therapy. The local concentration of conjugates near the tumor should be high, and any leakage from the cavity would result in a low systemic concentration. Thus, tumor cells should be readily killed while vital organs go unharmed. Also, the large size of the conjugate might slow the diffusion from the site of injection. A number of tumor models have been developed that can be used to evaluate intraperitoneal therapy of toxin conjugates.[43,56] Significant antitumor effects have been noted, and in one case, this has led to initiation of phase I clinical trials.[43]

Toxin conjugates also could be expected to have good access to target cells in the case of blood-borne malignancies such as leukemias and lymphomas. It has not been easy to establish therapeutic models of human leukemias in nude mice. However, treatment of a mouse B cell tumor, BCL_1, has shown some very encouraging results.[6] A small number of leukemia patients have been treated with immunotoxins.[57] Immunotoxins directed against human melanoma and human colon cancer have been administered to approximately 50 to 100 patients as part of a phase I/II clinical trial.[58] Thus, toxin conjugates are being evaluated as novel drugs and not just laboratory curiosities.

A serious drawback with giving conjugates in clinical setting is the strong immunogenic nature of the toxins. Universally, in both test animals and humans, reports indicate that a neutralizing antibody response develops against the toxin by days 10 to 14 after the first injection. If conjugates are to be effective, either immune tolerance protocols will have to be designed, or several different toxins used, or all therapeutic benefit will have to be achieved within a 2-week dosing period.

IX. EVOLUTION OF TOXIN STRUCTURE/FUNCTION

DT, PE, and ricin have each been cloned and sequenced.[40,59,60]. In addition, protein crystals of PE and ricin have been grown and analyzed. Such advances will help significantly in understanding how molecules, such as these toxins, have evolved.

Many bacterial toxins are actively secreted into the culture medium during growth. Later, when they kill cells, they translocate across a mammalian vesicle membrane. Are these two transport functions, the secretion and the translocation, related? The use of gene fusions and various conjugates should be of enormous use in answering such questions. DT and PE have the same enzymatic activity and yet they share little or no sequence homology and are unrelated immunologically. However, it should be possible to make toxin-toxin conjugates to determine if the various functions can complement each other. Some toxin-toxin conjugates have already been made between members of the ribosome-inactivating plant toxins. These conjugates have demonstrated that the A-chain from one toxin can function in place of the A-chain from a second toxin. However, the real value of such work may come when there is a better understanding of the location of the most and the least favorable regions of these toxins. It may be possible to make toxin-toxin gene fusions that will include only the favorable subregions.

Many bacterial toxins have ADP-ribosylating activity as a part of their cell-killing or cell-damaging activity. However, they have different substrates as acceptors for ADP-ribose. Again, by making toxin-toxin conjugates and gene fusions, we may be able to distinguish unique regions of each toxin which are involved in substrate recognition and common regions which are involved in NAD binding.

Can a single toxin molecule translocate out of endocytic vesicles and into the cell cytoplasm of mammalian cells or is there a need for cooperativity between toxin molecules, i.e., does there need to be a critical number of toxin molecules present to generate a lesion in an endocytic vesicle? A cooperative function has been proposed for the entry of DT.[61] To date, this cooperative function has not been assigned to any particular toxin domain. Presumably, techniques, such as molecular cloning and crystallographic analyses will help locate this and other toxin functions.

X. CONCLUSION

Toxin conjugates interact with mammalian cells in a variety of unique ways. These interactions reveal details about the nature of the toxins, the carrier proteins, and the target cells. Recent advances in molecular cloning and protein crystallography have made it possible to attribute various functions to specific toxin domains. A new and exciting potential use for toxin conjugates is in the area of clinical medicine.

ACKNOWLEDGMENTS

We would like to thank the following colleagues at the Laboratory of Molecular Biology for their valuable contributions to the work with PE: Mark Willingham, Sankar Adhya, Vijay Chaudhary, Haya Lorberboum-Galski, Toshi Kondo, and Yong-hua Xu. Our thanks also to Tom Waldmann and Robert Pirker. Technical help was supplied by Betty Lovelace, Maria Gallo, Angelina Rutherford, and Annie Harris. We thank Althea Gaddis for typing the manuscript and Susan Kane for editorial assistance.

REFERENCES

1. **Middlebrook, J. L. and Dorland, R. B.,** Bacterial toxins: cellular mechanisms of action, *Microbiol. Rev.,* 48, 199, 1984.
2. **Olsnes, S. and Sandvig, K.,** Entry of polypeptide toxins into mammalian cells, in *Endocytosis,* Pastan, I. and Willingham, M. C., Eds., Plenum Press, New York, 1985.
3. **Endo, Y. and Tsurugi, K.,** RNA N-glycosidase activity of ricin A-chain, *J. Biol. Chem.,* 262, 8128, 1987.

4. **Vitetta, E. S. and Uhr, J. W.**, Immunotoxins, *Annu. Rev. Immunol.*, 3, 197, 1985.

5. **Pastan, I., Willingham, M. C., and FitzGerald, D. J. P.**, Immunotoxins, *Cell*, 47, 641, 1986.

6. **Vitetta, E. S., Fulton, R. J., May, R. D., Till, M., and Uhr, J.**, Redesigning nature's poisons to create anti-tumor reagents, *Science*, 238, 1098, 1987.

7. **Pirker, R., FitzGerald, D. J. P., Willingham, M. C., and Pastan, I.**, Immunotoxins and endocytosis, *Lymphokines*, 14, 361, 1987.

8. **Blakely, D. C. and Thorpe, P. E.**, An overview of therapy with immunotoxins containing ricin or its A chain, *Antibody, Immunoconjugates and Radiopharmaceuticals*, 1, 1, 1988.

9. **Volkman, D. J., Ateeq, A., Fauci, A. A., and Neville, D. M., Jr.**, Selective abbrogation of antigen-specific human B cell responses by antigen-ricin conjugates, *J. Exp. Med.*, 156, 632, 1982.

10. **Allured, V., Collier, R. J., Carroll, S. F., and McKay, D. B.**, Structure of exotoxin A of *Pseudomonas aeruginosa* at 3.0-Angstrom resolution, *Proc. Natl. Acad. Sci. U.S.A.*, 83, 1320, 1986.

11. **Montfort, W., Villafranca, J. E., Monzingo, A. F., Ernst, S. R., Katzin, B., Rutenber, E., Xuong, N. H., Hamlin, R., and Robertus, J. D.**, The three-dimensional structure of ricin at 2.8 Å, *J. Biol. Chem.*, 262, 5398, 1987.

12. **Hwang, J., FitzGerald, D. J., Adhya, S., and Pastan, I.**, Functional domains of pseudomonas exotoxin identified by deletion analysis of the gene expressed in *E. coli*, *Cell*, 48, 129, 1987.

13. **Douglas, C. M. and Collier, R. J.**, Exotoxin A of *Pseudomonas aeruginosa*: substitution of glutamic acid 553 with aspartic acid drastically reduces toxicity and enzymatic activity, *J. Bacteriol.*, 169, 4967, 1987.

14. **Douglas, C. M., Guidi-Rontani, C., and Collier, R. J.**, Exotoxin A of *Pseudomonas aeruginosa*: active, cloned toxin is secreted into the periplasmic space of *Escherichia coli*, *J. Bacteriol.*, 169, 4962, 1987.

15. **Chaudhary, V. K., Xu, Y.-h., FitzGerald, D., Adhya, S., and Pastan, I.**, Role of domain II of pseudomonas exotoxin in the secretion of proteins into the periplasm and medium by *E. coli*, *Proc. Natl. Acad. Sci. U.S.A.*, 85, 2939, 1988.

16. **FitzGerald, D. J. P., Padmanabhan, R., Pastan, I., and Willingham, M. C.**, Adenovirus-induced release of epidermal growth factor and pseudomonas toxin into the cytosol of KB cells during receptor-mediated endocytosis, *Cell*, 32, 607, 1983.

17. **Cumber, J. A., Forrester, J. A., Foxwell, B. M. J., Ross, W. C. J., and Thorpe, P. E.**, Preparation antibody-toxin conjugates, *Methods Enzymol.*, 112, 207, 1985.

18. **FitzGerald, D. J. P.**, Construction of immunotoxins using Pseudomonas exotoxin A, *Methods Enzymol.*, 151, 139, 1987.

19. **Colombatti, M., Greenfield, L., and Youle, R. J.**, Cloned fragment of diphtheria toxin linked to T cell-specific antibody identifies regions of B chain active in cell entry, *J. Biol. Chem.*, 261, 3030, 1986.

20. **Greenfield, L., Johnson, V. G., and Youle, R. J.**, Mutations in diphtheria toxin separate binding from entry and amplify immunotoxin selectivity, *Science*, 238, 536, 1987.

21. **Kondo, T., FitzGerald, D., Chaudhary, V. K., Adhya, S., and Pastan, I.**, Activity of immunotoxins constructed with modified pseudomonas exotoxin A lacking the cell recognition domain, *J. Biol. Chem.*, 263, 9470, 1988.

22. **Bjorn, M. J., Groetsema, G., and Scalapino, L.**, Antibody-pseudomonas exotoxin A conjugates cytotoxic to human breast cancer cells *in vitro*, *Cancer Res.*, 46, 3262, 1986.

23. **Thorpe, P. E., Wallace, P. M., Knowles, P. P., Relf, M. G., Brown, A. N. F., Watson, G. J., Kryta, R. E., Wawagynczak, E. J., and Blakely, D. C.**, New coupling agents for the synthesis of immunotoxins containing a hindered disulfide bond with improved stability, *Cancer Res.*, 47, 5924, 1987.

24. **Blattler, W. A., Kuenzi, B. S., Lambert, J. M., and Senter, P. D.**, New heterobifunctional protein cross-linking reagent that forms an acid-labile link, *Biochemistry*, 24, 1517, 1985.

25. **Murphy, J. R., Bishai, W., Borowski, M., Miyanohara, A., Boyd, J., and Nagle, S.**, Genetic construction, expression and melanoma-selective cytotoxicity of a diphtheria toxin-related alpha-melanocyte-stimulating hormone fusion protein, *Proc. Natl. Acad. Sci. U.S.A.*, 83, 8258, 1986.

26. **Chaudhary, V. K., FitzGerald, D. J., Adhya, S., and Pastan, I.**, Activity of a recombinant fusion protein between transforming growth factor type alpha and Pseudomonas toxin, *Proc. Natl. Acad. Sci. U.S.A.*, 84, 4538, 1987.

27. **Lorberboum-Galski, H., FitzGerald, D. J. P., Adhya, S., and Pastan, I.**, Cytotoxic activity of an interleukin 2-Pseudomonas exotoxin chimeric protein produced in *E. coli*, *Proc. Natl. Acad. Sci. U.S.A.*, 85, 1922, 1988.

28. **Pappenheimer, A. M.**, Diphtheria toxin, *Annu. Rev. of Biochem.*, 46, 69, 1977.

29. **Laird, W., and Groman, N.**, Isolation and characterization of *tox* mutants of corynebacteriophage beta, *J. Virol.*, 19, 220, 1976.

30. **Neville, D. M., Jr. and Youle, R. J.**, Monoclonal antibody-ricin or ricin A chain hybrids: kinetic analysis of cell killing for tumor therapy, *Immunol. Rev.*, 62, 75, 1982.

31. **FitzGerald, D. J., Bjorn, M. J., Ferris, R. J., Winkelhake, J. L., Frankel, A. E., Hamilton, T. C., Ozols, R. F., Willingham, M. C., and Pastan, I.**, Antitumor activity of an immunotoxin in a nude mouse model of human ovarian cancer, *Cancer Res.*, 47, 1407, 1987.

32. **Cawley, D. B. and Herschman, H. R.,** Epidermal growth factor-toxin A chain conjugates: EGF-ricin A is a protein toxin while EGF-diphtheria fragment A is nontoxic, *Cell,* 22, 563, 1980.

33. **Raso, V. and Basala, M.,** A highly cytotoxic human transferrin-ricin A chain conjugate used to select receptor-modified cells, *J. Biol. Chem.,* 259, 1143, 1984.

34. **Trowbridge, I. S. and Domingo, L. D.,** Anti-transferrin receptor monoclonal antibody and toxin-antibody conjugates affect growth of human tumor cells, *Nature,* 294, 171, 1981.

35. **Kronke, M., Depper, J. M., Leonard, W. J., Vitetta, E. S., Waldmann, T. A., and Greene, W. C.,** Adult T cell leukemia: A potential target for ricin A chain immunotoxins, *Blood,* 65, 1416, 1985.

36. **Cassellas, P., Bourrie, B. J. P., Gross, P., and Jansen, F.,** Kinetics of cytotoxicity induced by immunotoxins: enhancement by lysosomotropic amines and carboxylic ionophores, *J. Biol. Chem.,* 259, 9359, 1984.

37. **Ramakrishnan, S. and Houston, L. L.,** Inhibition of human acute lymphoblastic leukemia cells by immunotoxins: potentiation by chloroquine, *Science,* 223, 58, 1984.

38. **Thorpe, P. E., Ross, W. C. J., Brown, A. N. F., Myers, C. D., Cumber, A. J., Foxwell, B. M. J., and Forrester, J. T.,** Blockade of the galactose-binding sites of ricin by its linkage to antibody, *Eur. J. Biochem.,* 140, 63, 1984.

39. **Weil-Hillman, G., Runge, W., Jansen, F. K., and Vallera, D. A.,** Cytotoxic effect of anti-M_r 67,000 protein immunotoxins on human tumors in a nude mouse model, *Cancer Res.,* 45, 1328, 1985.

40. **Gray, G. L., Smith, O. H., Baldridge, J. S., Hoskins, R. N., Vasil, M. L., Chen, E. Y., and Heynelcer, H. L.,** Cloning, nucleotide sequence and expression in *Escherichia coli* of the exotoxin A structural gene of *Pseudomonas aeruginosa, Proc. Natl. Acad. Sci. U.S.A.,* 81, 2645, 1984.

41. **Pirker, R., FitzGerald, D. J. P., Hamilton, T. C., Ozols, R. F., Laird, W., Frankel, A. E., Willingham, M. C., and Pastan, I.,** Characterization of immunotoxins active against ovarian cancer cell lines, *J. Clin. Invest.,* 76, 1261, 1985.

42. **FitzGerald, D. J., Willingham, M. C., and Pastan, I.,** Antitumor effects of an immunotoxin made with pseudomonas exotoxin in a nude mouse model of human ovarian cancer, *Proc. Natl. Acad. Sci. U.S.A.,* 83, 6627, 1986.

43. **Willingham, M. C., FitzGerald, D., and Pastan,I.,** Pseudomonas exotoxin coupled to a monoclonal antibody against ovarian cancer inhibits the growth of human ovarian cancer cells in a mouse model, *Proc. Natl. Acad. Sci. U.S.A.,* 84, 2474, 1987.

44. **FitzGerald, D. J. P., Waldmann, T. A., Willingham, M. C., and Pastan, I.,** Pseudomonas exotoxin-Anti-Tac: Cell specific immunotoxin active against cells expressing the human T cell growth factor receptor, *J. Clin. Invest.,* 74, 966, 1984.

45. **FitzGerald, D. J. P., Trowbridge, I. S., Pastan, I., and Willingham, M. C.,** Enhancement of toxicity of antitransferrin receptor antibody-pseudomonas exotoxin conjugates by adenovirus, *Proc. Natl. Acad. Sci. U.S.A.,* 80, 4134, 1983.

46. **Pirker, R., FitzGerald, D. J. P., Hamilton, T. C., Ozols, R. F., Willingham, M. C., and Pastan, I.,** Anti-transferrin receptor antibody linked to pseudomonas exotoxin as a model immunotoxin in human ovarian carcinoma cell lines, *Cancer Res.,* 45, 751, 1985.

47. **Bjorn, M. J., Groetsema, G.,** Immunotoxins to the murine transferrin receptor: intracavitary therapy of mice bearing syngeneic peritoneal tumors, *Cancer Res.,* 47, 6639, 1987.

48. **Cawley, D. B., Simpson, D. L., and Herschman, H. R.,** Asialoglycoprotein receptor mediates the toxic effects of an asialofatuin-diphtheria toxin fragment A conjugate on cultured rat hepatocytes, *Proc. Natl. Acad. Sci. U.S.A.,* 78, 3383, 1981.

49. **Youle, R. J., Murray, G. J., and Neville, D. M., Jr.,** Studies on the galactose-binding site of ricin and the hybrid toxin Man6P-ricin, *Cell,* 23, 551, 1981.

50. **Levin, L. V., Griffin, T. W., Childs, L. R., Davis, S., and Haagensen, D. E., Jr.,** Comparison of multiple anti-CEA immunotoxins active against human adenocarcinoma cells, *Cancer Immunol.,* 24, 202, 1987.

51. **Uckum, F. M., Jaszcz, W., Ambrus, J. L., Fauci, A. S., Gajl-Peczalska, K., Song, C. W., Wick, M. R., Myers, D. E., Waddick, K., and Ledbetter, J. A.,** Detailed studies on expression and function of CD19 surface determinant by using B43 monoclonal antibody and the clinical potential of anti-CD19 immunotoxins, *Blood,* 71, 13, 1988.

52. **Lyall, R. M., Hwang, J., Cardarelli, C., FitzGerald, D., Akiyama, S., Gottesman, M. M., and Pastan, I.,** Isolation of human KB cell lines resistant to epidermal growth factor-Pseudomonas exotoxin conjugates, *Cancer Res.,* 47, 2961, 1987.

53. **Goldmacher, V. S., Anderson, J., Schulz, M. L., Blattler, W. A., and Lambert, J. M.,** Somatic cell mutants resistant to ricin, diphtheria toxin and to immunotoxins, *J. Biol. Chem.,* 262, 3205, 1987.

54. **Velu, J. J., Beguinot, L., Vass, W. C., Willingham, M. C., Merlino, G. T., Pastan, I., and Lowy, D. R.,** Epidermal growth factor-dependent transformation by a human EGF receptor proto-oncogene, *Science,* 238, 1408, 1987.

55. **Carney, D. N., Cuttitta, F., Moody, T. W., and Munna, J. D.,** Selective stimulation of small cell lung cancer clonal growth by bombesin and gastrin-releasing peptide, *Cancer Res.,* 47, 821, 1987.

56. **Griffin, T. W., Richardson, C., Houston, L. L., Le Page, D., Bogden, A., and Ross, V.,** Antitumor activity of intraperitoneal immunotoxins in a nude mouse model of human malignant mesothelioma, *Cancer Res.,* 47, 4266, 1987.

57. **Laurent, G., Pris, J., Farcet, J. P., Carayon, P., Blythman, H., Cassellas, P., Poncelet, P., and Tansen, F. K.,** Effects of therapy with T101 ricin A-chain immunotoxin in two leukemia patients, *Blood,* 67, 1680, 1986.

58. **Spitler, L.,** Immunotoxin therapy of malignant melanoma, *Med. Oncol. Tumor Pharmacother.,* 3, 147, 1986.

59. **Greenfield, L., Bjorn, M. J., Horn, G., Fong, D., Buck, G. A., Collier, R. J., and Kaplan, D. A.,** Nucleotide sequence of the structural gene for diphtheria toxin carried by corynebacteriophage beta, *Proc. Natl. Acad. Sci. U.S.A.,* 80, 6853, 1986.

60. **Lamb, F. I., Roberts, L. M., and Lord, J. M.,** Nucleotide sequence of cloned cDNA coding for preproricin, *Eur. J. Biochem.,* 148, 265, 1985.

61. **Hudson, T. and Neville, D. M., Jr.,** Quantal entry of diphtheria toxin to the cytosol, *J. Biol. Chem.,* 260, 2675, 1985.

INDEX